W9-CLA-358

FIELD,
FLOWER,
VASE

FIELD, FLOWER, VASE

ARRANGING *and* **CRAFTING**

WITH

SEASONAL *and* **WILD BLOOMS**

BY

CHELSEA FUSS

ADDITIONAL PORTRAITS AND PHOTOS BY

CLÁUDIO SILVA

A summer meadow of poppies and corn marigold atop a small mountain near Sintra, Portugal.

For my parents,
Michael and Dorothyanne,
who always encouraged
my love for flowers.

PART II

FLOWERS BY ROOM

1—Living Room

2—Kitchen

3—Bedroom

4—Bath

5—Garden and Terrace

The livestock trail in Sintra where I forage.

Preface

THE IDEA OF FORAGING—WHETHER FOR EDIBLE PLANTS, dye plants, or just decorative ones—has had a resurgence in recent years. But what is it? Isn't it just a funny excuse for stealing your neighbor's flowers?

Foraging is, at its essence, about searching. Searching for herbs, berries, blossoms, and fruit for decorating or eating. Perhaps it is the search that makes this practice so intriguing in our modern world, forcing us to slow down and notice every nook and cranny as we scour the landscape for ingredients.

Searching was exactly what I was doing when I left a settled life in the United States to traipse around Europe with a backpack, working on organic farms for ten months, at an age that seemed too old to be doing such things. Looking back, I remember searching for flowers and plants all along my travel route, among all the other things I seemed to be searching for involving every cliché of home, community, love, and self. I gathered bouquets wherever I went, whether it was saving rose blossoms from the compost on a Brittany farm, scouring a Gotland forest for autumn berries, or clipping olive branches on the outskirts of Beja, Portugal.

In my search for home, I eventually found Lisbon and after a false start at settling, relocated to a small village outside of Sintra, a forty-minute train ride from Lisbon and a place Lord Byron called "the most beautiful village in the world." But it was more than that. Always a gathering place for artists and mystics, Sintra holds intrigue with its microclimate, encouraging vegetation of all varieties to prosper there year-round, resulting in verdant hillsides accented with crumbling stone walls and abandoned villas that feel worlds away from the intensity of Lisbon.

In a small hamlet outside of the village, I found a plaster-walled, cottage-like home inside of a little cluster of stucco dwellings by a creek, with a large meadow in the back. I secured a garden plot in the idyllic pasture after negotiating with the landlord, making a promise to maintain an organic garden on the property as her grandmother had always done.

An abandoned villa overgrown with ivy and Virginia creeper.

It was in that rustic little cottage and garden where I turned to flowers for meditation, play, and a sense of grounding while so far from home. Though I've worked with flowers as a florist and stylist for over twenty years, this relationship with flowers was different. Foraging for wild stems offered the healing that I'd been craving during that time.

Whether in the garden harvesting tiny strawberry flowers, out foraging elderflower on the trail, making simple handheld bouquets from sidewalk-crack weeds, arranging discarded tomato branches for my entry, in the kitchen concocting new recipes with homegrown marigolds, or creating restful rituals with rose petals, I began to cultivate a new life with flowers at the center. I fell in love with a trail where I would lose myself on my hikes in and out of the village, searching for small stems of Spanish daisies, vines of passionflower, Queen Anne's lace, and wild grasses. I brought them home for arrangements, woven wreaths, and recipes, all made with what I had available right there in the village. This book is a way to share this unhurried, floral-focused life with you with the hope that you'll be inspired to create your own flower-filled days.

Gathering greenery in a vintage Portuguese basket.

Freshly foraged wild corn marigold, oats, poppy, wild rose, and false dandelion.

Introduction

IN THIS BOOK, YOU WILL FIND ways to tune into your truest instincts with flowers and to bring floral fragrance and beauty into every room of your home using all-natural ingredients and sustainable techniques. At its heart, this is a guidebook aimed at encouraging you to bring more flowers into your daily life and to immerse yourself in the joy and healing nature of fresh blossoms, herbs, stems, leaves, branches, twigs, and berries.

To create a life where flowers are part of the everyday does not require any sort of budget, access, or travel. Flowers are not just for special occasions or for the wealthy: They can be casual, local, unfussy—and they can be found everywhere, from your backyard or your neighborhood debris pile to your local, old-school flower shop or grocery store.

The book you hold in your hands was born from the workshops of the same name, which I began holding in Sintra after having my own sort of aha moment with foraging. In my workshops, I take tourists and locals out on favorite trails, and even to a secret, nearly abandoned villa adjacent a forest, to search for materials and then use the seasonal ingredients we find. My students learn through instinct and by doing, using the meditative process of gathering and using what is available. I often hear that students continue this pleasant practice at home, months later, plucking stray blossoms, scouring their neighborhoods for weeds, and continuing the habit of gathering floral ingredients for wellness, for their homes, and for a bit of floral therapy each week.

Because resources were scarce compared to the plentiful floral suppliers I was used to when I first moved to Portugal, I developed new floral-arranging techniques that were more sustainable for the environment, local economies, and my own pocketbook. I turned to natural twines instead of wire, grapevines instead of steel wreath frames, and stones instead of flower pin holders. And as a result, I created work that felt more at home in its environment and more environmentally friendly.

A wild nasturtium growing
in a forest near Sintra.

Like my workshops, this book takes a very natural approach to floral design. The flower varieties I use are common throughout many types of climates, but, of course, the idea is to source from your own neighborhood and local shops, taking liberties with substitutions as you like. I've included many herbs in the projects—part of my flower-arranging process since I was a child—which I use for their culinary purposes, blossoms, and benefits for health and well-being.

Once I found the cadence of the landscape and seasons in my small village, everything else I'd made before felt a bit contrived and the new floral creations began to flow in an effortless sort of way. I am convinced anyone can find the pulse of their local landscape and culture to create flower arrangements that grow literally and visually from their surroundings.

In the first four chapters, you will learn all the basics you need to begin crafting your own projects with fresh flowers, including how to approach flowers like a stylist for your own home. Next, we will explore every room of the house, with flower ideas for each, including arrangements for spontaneous celebrations, flowers that help you sleep, edible flowers, and flowers for everyday decorating.

Each project is informed by the raw landscape of Portugal. Nothing too manicured, and always embracing the fleeting lifespan of just-picked wildflowers. This mind-set can be implemented anywhere, and I hope the projects in this book encourage you to adopt it. If you find flowers close to home, and in the process start breathing a little slower and stepping away from the screen a bit more often, you've found the magic of flowers I've hoped to share.

part I

HOW TO WORK *with* CUT FLOWERS

Chicory, willowherb, and dandelion growing by a roadside.

LET THE FLOWERS LEAD THE WAY

I prefer to think of flower arranging as taking a small sampling of nature inside; a few blooms from the corner of a pasture, a pocket of weeds, or an overhanging vine found on a walk adds life and vibrancy to any room.

There are no right or wrong ways to arrange flowers if you work as instinctively as possible, tune into yourself, and gather the ingredients that speak to you. I think flowers should be playful and fun and never too serious. I don't want to live with formal branches in urns but with wild stems of daisies in drinking glasses, imperfect climbing roses, and hydrangeas in pudding bowls. I want my flowers untamed, unmanicured.

Dandelions growing by a doorstep in Colares, Portugal.

Rarely planning out an arrangement from start to finish, rather—much like how one might shop the farmers' market for produce in the summer—I let myself be inspired by what is in season. And when creating a composition, I don't always set out with an expected shape or color scheme; instead I follow the curve of a vine and nature's color palette of the moment. I err on the side of unruly and primitive, as opposed to symmetrical and manicured. I like to capture the essence of a freestyle flower meadow in a vase by seeking out flowers that are soft and blowsy, even when shopping at a grocery store.

Forget everything you think a flower arrangement should look like. Don't be afraid for flowers to overlap as they do in nature, to wilt, to dry, or to change, embracing their wild state. Find out what your local landscape produces best, as well as the specialties of your local markets and shops. Go your own way, with the materials around you.

Spanish daisies, wild roses, grasses, and dandelion seed heads mimic a summer wildflower meadow.

Making wildflower arrangements in early summer.

Dandelion seed heads mixed with wildflowers.

A honeysuckle vine leads the way for a summer composition of wildflowers in a drinking glass.

Honeysuckle, mustard, and oat grass offer a blowsy sampling of nature.

Climbing roses and wisteria being processed in a sink.

Privet in bloom by a roadside

A meadow of wild mallow and cow parsley in Colares, Portugal.

A pitcher of wild roses creates a corner of nature on a windowsill.

THINK LIKE A STYLIST

I think more like a stylist than a florist when working with flowers, focusing on how the blooms add to the existing story of a room instead of making them the center of attention in a grand way. Formal flowers that call attention to themselves are perfectly beautiful in the right context, but in most of our homes they would feel like museum pieces that can't be touched. Instead, finding ways to include bits of imperfection in our everyday world feels more natural.

A pitcher of summer wildflowers, gathered at all stages of life, adds depth to a corner of a room, creating a dynamic element in the interior landscape. When flowers are nestled organically into the design of the room, they feel like an essential element of the household.

Think about the way Matisse used flowers in his still life artwork. The flowers were part of a larger vignette, playing a role in a larger scene of mixed textiles and knickknacks, and not at all too preciously placed.

Wild thistle in Penedo, Portugal.

CONSIDER CONTEXT

In addition to starting with local flora, context is a key element of a floral arrangement and the foundation from which every other element will naturally grow. Let your current home, tastes, and local specialties inspire the style and composition of your arrangements. Each flower has its own personality and mood; find what is right for your space. Take cues from the wall colors, furniture, and collections you already own, and consider that most likely the best flowers for your space are the ones you can find within a short walk from home.

Once I tried to use store-bought ranunculus in my country cottage, but they felt too domesticated for my rustic space. It turns out, the best flowers for my plaster-walled, crooked-floored house were the ones I grew or could find right in my neighborhood.

Use containers that are already in your space—drinking glasses, cereal bowls, grandma's tea jar, ceramics that you already love—and fill them with the flowers that you are drawn to. Even a single stem of grass can make a difference to the energy of a room.

I also think about context in terms of how I've seen flowers used. And this can be very personal. You will have your own associations for flowers. For instance, alstroemeria is a flower that I grew up seeing at the wholesale market and at grocery stores, covered in plastic sleeves and beloved for its long vase life. For me, it will always symbolize a grocery store, and I just can't get beyond it. On the other hand, a branch of lilac will always remind me of my childhood gardens in New Hampshire, where I instinctively gathered branches in the rain for primitive, unconsidered bouquets. But for you, each of these flowers might have a different association.

Wild fennel and oat grass feel at home in a country cottage in Colares, Portugal.

Morning glory growing in Penedo, Portugal.

EMBRACE LIFESPAN

I like to experience the full lifespan of flowers as much as possible. To me, flowers are what pets are to some other people—they keep me company, changing every day and bringing a dynamic piece of life to my space. Because of this, I like the idea of embracing their changing forms over a long vase life.

I recently made a small arrangement with a morning glory vine, a plant many consider to be invasive and problematic in their gardens. (I see it growing in parking lots and nearly covering abandoned buildings, so for me, it is a reminder of nature overtaking an urban landscape.)

Since morning glory flowers only last a day, I plucked each blossom off the vine once it was spent, and each morning a new one appeared. In total the vine lasted only about three days in a warm room, but the experience was interactive, full of life, change, and death. What a great reminder to adjust our expectations of cut flowers and accept flowers as they naturally are.

Morning glory in my windowsill.

A branch from a strawberry tree styled in a traditional pitcher from Spain.

NO MATERIALS ARE OFF-LIMITS

For me, a dandelion holds us much intrigue as a rose. When foraging, think about the materials you might overlook at a flower shop. Instead of looking for the brightest colors or most showy blossoms, you might be satisfied with any bit of color, texture, interesting scale, or materials that surprise. It's not unlike the idea of wanting what you have in order to appreciate abundance in your life. That's the meditative part of it, too. Perhaps it is October and there are no flowers, but try to look for different shades of green, contrasting leaf sizes, small buds as texture, and seedpods for surprising shapes.

LOOKING TO NATURE'S COMPOSITIONS

First, look at the meadows. If you live in a city, look at the way the flowers are growing in an abandoned parking lot. How is nature taking over when left unchecked? Look to the unmanicured areas, not the domesticated spaces or designed gardens.

You'll likely notice that flowers almost always grow in clumps, even in wild areas. You see several blossoms all together extending from one plant. This is where I get a sense of repetition in my arrangements, though there's a fine line between repetition and looking like you just tossed a dozen roses into a vase. While wildflowers like mustard or poppies can grow in profuse quantities, a flower like a rose sometimes grows more sparsely in less cultivated areas, so if you have twenty-four stems of the same rose, the ingredients are obviously purchased and commercially grown. In the wild we do have repetition, but it is alongside a great variety of plants. The beauty of working with found flowers is that often you will have intense texture from so many varieties. It's like when you grow a garden at home. You might have a couple of tomatoes, a few stems of basil, a few stems of this and that to harvest. You won't have the huge amounts that we have in the supermarket. I think our eyes were trained in America to see a certain type of abundance in the overindulgence of overflowing bakeries, flower shops, and large shops with endless selections, when in fact a few thoughtful items can be just as special.

I like to see the different heights in nature, and to notice what is growing together. Often what grows together, I put in arrangements together. If all else fails, copy nature!

A summer wildflower hill in
Penedo, Portugal.

Wild nasturtium growing in a valley.

Where do you find wild, blowsy flowers? Where do you forage? And if foraging for flowers isn't possible, where do you buy them? I've heard so often from people that it's not possible for them to buy flowers because all of the flower shops in their towns are so limited, or they don't have a meadow to forage in. I bet that, in fact, you have some options.

Whenever I am traveling, I seek out flower shops of any variety, just like I do weeds in the sidewalk cracks or meadows where I can clip ingredients for floral compositions. While I have my own specific tastes, I've learned to search out possibilities wherever I am. And I've developed an increasing appreciation for the strong and steady florists still operating shops after thirty years. They may not be the trendiest, but they are flower-lovers working hard in our community. Developing a relationship with them can in turn produce some good results in terms of sourcing ingredients. Similarly, developing relationships with your local farmers, neighbors, and wholesale flower sellers can help you stay open to all the options.

Gathering greens in a basket.

A vintage Portuguese basket I use to gather overhanging geranium, ivy, and greens.

Nasturtium and acanthus growing wild next to a parking lot.

A traditional Portuguese horta garden.

Foraging near some abandoned buildings in Lisbon.

FORAGING

Look in abandoned lots, traffic islands, parking lots, and wild areas that are unkept in your city. You should ask for permission before taking anything from private property. Be wary of public parks, protected areas, and particularly protected wildflowers. Some areas have laws against picking wildflowers. If you aren't sure, don't pick—or check with whomever is in charge first. The positive side of foraging is that in many cases, clipping flowers can help them keep blooming, or taking out invasive vines like ivy can assist the local vegetation in thriving. Many areas clear invasive plant material in the summer to avoid fires, so foraging this material can assist in that practice. Only take a small portion of one plant, unless it's invasive. National and regional parks usually have their own rules about what you can and can't forage, so find out what the rules are before you clip.

It also pays to be aware of common poisonous plants in your area. Listed at right are some common wild plants you might encounter that are poisonous and can even cause harmful reactions on contact (see Safety, page 238).

When foraging, be careful about foraging low to the ground where runoff or animal waste can be present (since the plants may touch your skin or household surfaces). I try to always forage at waist level or higher, and well away from roads. Make sure you only forage items that you can identify with 100 percent certainty. Even if you are foraging for ingredients for arrangements, there are varieties that can cause skin irritations and even make you sick. I try to always cut varieties that I can identify, and I try to find out about plants that I don't recognize. That said, there is a lot of amazing material out there that will add the perfect meadow appeal to your arrangements. Go for a walk in your neighborhood and see what you can find.

FAVORITE FORAGING SPOTS

- Overgrown lots
- Parking lots that are unmanicured
- Neighborhood garden debris piles
- Neighbors (If you see something you want, knock on the door and ask if you can buy a few stems.)
- Public paths and hiking trails
- Roadsides
- Landscapers (Talk to landscapers you see working, and ask if you can have some of the branches they are throwing out from their pruning work.)

COMMON POISONOUS PLANTS

- Hogweed
- Lantana
- Oleander
- Poison hemlock
- Poison ivy
- Poison oak
- Poison sumac
- Water hemlock

GREAT OPTIONS FROM THE PRODUCE AISLE

- Chives
- Dill
- Fennel
- Kale
- Mint
- Parsley
- Rosemary

AT THE GROCERY STORE

Many grocers also have florists you can talk to about where they get their flowers and about how you would like to see locally sourced flowers. Solidago and statice, which are found at almost every grocery store in America, are easy to layer in arrangements for a wispy, ethereal feel. You can also look past the florist section and into the produce and plant aisles. I love to include blooming herbs and blooming veggies, like kale and radish, in arrangements. I also buy the herb plants they sell in the grocery store, cutting some for a bouquet and then replanting the rest in the garden.

HOW TO SHOP THE FARMERS' MARKET

- Ask veggie growers if they have any flowers
- Order flowers ahead
- Ask questions!
- Request varieties you want

FARMERS' MARKET

One of my favorite sources of local flowers is the farmers' market, and not just the flower stall! Start up conversations with the veggie sellers. Ask if they have any flower patches at their farm. I recently asked a produce grower if he had any apples on the branch, and the next week he brought an entire box for me. They often have little plots of flowers that they think nothing of, being farmers, but that are gold to the florally minded. I have two lovely ladies at my local farmers' market that I buy edible flowers from, and often they have more back on the farm, so I just need to ask if I want more of one variety.

Buying locally grown statice from a seller in Almoçageme, Portugal

Apple branches sourced from a fruit seller at Lisbon's organic produce market in Principe Real.

Herbs and other plants at a market in Almoçageme, Portugal.

Corn marigold growing in the wild.

On a walk in the neighborhood of São Pedro in Sintra, looking for materials and sources.

Statice and star of Bethlehem at a farmers' market.

A small market in Colares, Portugal, sells statice alongside fresh produce.

Clover growing in a sidewalk crack.

Blooming herbs collected from a farmers' market to use in arrangements.

WHOLESALERS

When I had a flower shop, I bought from farmers who had a stall at my local wholesale market, and sometimes I visited their farms, too. Generally, every big city has a wholesale flower market. This is where flowers are flown in from all around the world or where local growers sell their stems. In my home city of Portland, Oregon, we have a large local farm section, and then three or four other large wholesalers that bring in materials from around the world.

It is quite a process to get certified to go shop in this market, and you must prove that you are operating a business. But it works differently in every city. In some cities, the wholesale market has public hours and private hours. Research your local wholesalers and consult the list of wholesalers in the back of this book. At wholesalers, all the stems are sold in bulk, generally twenty-four stems to a bunch for items like roses or ten to fifteen per bunch for garden stems and fillers. The great thing about wholesalers is that they will often track down a particular variety for you, and they have access to foragers who bring in items that you might not have access to as well. In Lisbon, there is an older gentlemen who forages violets in the winter and brings them to our main wholesaler, who then resells them to all the flower shops.

Learning about your local wholesaler is also important if you have any interest in working with flowers on a commercial level. Wholesalers also sell at lower prices because they are selling in bulk to businesses who will resell the product.

Quince and viburnum from a local wholesaler.

A selection of fresh flowers from my local wholesaler and farmers' market.

FLOWER SHOPS

While it's nice to have access to wholesalers for the variety and larger quantities, if you are buying for home, you rarely need to purchase flowers in bulk. Instead, it makes more sense to develop a relationship with your local florist. When I started working with what I had locally and shopping in local flower shops, I learned that the shop owners could get local crops for me. I also learned that they often carried foraged items. My local florist, Florista Camélia in Sintra (next page), sometimes has foraged camellia, particularly in the spring and during Sintra's camellia festival. You almost always see camellia greens outside her shop. The camellia has a historical link to Sintra: In the 1800s, camellias were introduced to the town by King Ferdinand and have been celebrated with a yearly festival ever since. There are camellias planted on estates and farms all over Sintra, and their flowers bloom all winter. So, for locals, this shop has special meaning—and you may find a similar treasure in your own neighborhood.

FLOWER FARMS

While many flower farms are private businesses, closed to the public, there are just as many that allow customers to come and cut or buy precut bunches. Check out your local options! I've been to flower-cutting farms all over the world from small towns in the United States to Germany, Sweden, and Portugal. Most communities have some sort of flower farm where you can access freshly grown flowers (check the Resource and Shopping Guide in the back of this book and you may discover a place you never knew existed!).

My local florist, Amália, at her shop, Florista Camélia.

HOW TO SHOP FOR FLOWERS

When I go into a flower shop, I look for the soft flowers—nothing too stiff or upright—or flowers that might pair well with foraged items. Keeping to a single color sometimes makes it easier to create an arrangement, especially if you are working with a limited selection.

Where grocery store flowers often go wrong is in their scale. I see people bring home a mixed bouquet and just stick it straight into a vase. Often, the large flowers, like lilies, feel way out of scale, and all the flowers sit at the same height or in a stiff triangular shape. It is best to avoid the mixed-flower bouquets, and to instead shop single varieties of flowers in order to make your own mix. When you bring them home and place them in a vase, cut some of the flowers down so that they sit at the rim of the vase. Other lighter, smaller-scale flowers can be left at a taller height. If a mixed bouquet is your only option, you can separate the varieties out and mix them with garden or foraged flowers. When options are limited, sticking with one color will create more visual impact.

A magnolia just beginning to open.

HOW TO TELL IF FLOWERS ARE FRESH

The more you shop for flowers, the more you'll understand how to recognize freshness. Start at the stem and look for firm green leaves. If you see yellow leaves toward the bottom or detect any scent of rotting, the flowers have been sitting at the shop too long! Next, if the blossoms are all the way open, the days are numbered for the stem. That said, I often buy open blossoms and mix them with just-opening buds because I love to see flowers in different stages within one arrangement. As the arrangement ages, I just pick out the dead stems and enjoy the progression. It helps to think about shopping for flowers the same way you shop for your vegetables: Check color, texture, and scent, and look for stems that were not picked too early and have not become too "ripe" in the store. Often at groceries or large flower shops, you'll see red roses that are quite budded, showing no signs of opening, which means they were picked too early (and then transported across oceans and kept in the florist's fridge to look fresh upon arrival). Once the buds reach the normal temperature of your home, they will no longer be able to open and they will just fail right away. So here are a few things to check when assessing floral freshness.

FRESH FLOWER CHECKLIST

- Buds just opening (not too closed, and not all the way open)
- Good color in blossoms and leaves
- Firm, healthy leaves (not drooping)
- No browning or yellowing at the edges of leaves and petals
- No excessive petal or bud loss when you pick up the bouquet
- Flowers that smell fresh (no decaying scent)

HOW TO GROW FLOWERS

It's quite easy to grow flowers for cutting whether you have a tiny balcony or a full garden plot. You can plant a simple cutting garden in an afternoon; and if you are already growing vegetables, mixing them among the rows is a great strategy.

In Portugal, there are many community gardens; some are legal and others are renegade endeavors (which I find quite beautiful). In Portuguese, a cutting garden or kitchen garden is called a ***horta***, and you see them everywhere, even in places along the freeway. These gardeners mix in everything from vegetables to citrus to flowers. Their designs are similar to English cottage gardens or French ***potagers***, mixing herbs, vegetables, and flowers for a pragmatic but visually stunning little plot. Later in the book (see page 211) you will find instructions for growing a cutting garden on your terrace.

Early summer in a flower garden in Penedo, Portugal.

A small garden of homegrown poppies, nasturtium, scented geranium, and honeysuckle.

Twine and my favorite Japanese clippers.

My (nearly) all-natural toolbox is quite simple. I've always been a bit of a minimalist with floral tools, and I have a lot of respect for the classical methods of wiring and creating large installations, but as a home floral designer, you really don't need much! In recent years, I've moved completely away from floral foam (I used to occasionally use it for events) and coated wires, and I am moving away from plastic, which I once used to line vases that were not watertight. I sometimes use flower pins, also called flower frogs, but not very often, since I find they inhibit the movement I aspire to create in an arrangement.

I've adopted more renewable resources like glass, copper, and very simple natural elements like stones, twine, sticks, and branches to create floral pins. I love creating my own flower frogs with what is available to me and credit the florists Tom Pritchard and Billy Jarecki, owners of Madderlake, a celebrated New York City flower shop in the 1970s—and the brilliant flower books they created in the eighties and nineties—for showing me some of my favorite methods. I often use twigs wrapped in a vase, or a strategic part of a branch with a side shoot that creates a little nook, to hold stems. I also try to buy locally made clippers or used tools and to take care of the ones I have. Since I work in a very humid climate, I sometimes have a problem with rust on my clippers, so I scrub them with a bit of vinegar and dry them in the sun every few months.

It's important to find tools that work for you. A list of a few of my essentials follows.

Baskets, shown here in a variety of styles at a local market, are indispensable for a forager.

FAVORITE TOOLS

JAPANESE BONSAI CLIPPERS
I find these to be the most comfortable for my hands.

STONES AND PEBBLES
For keeping stems in place

TWINE
All widths. I like a narrow linen twine, and use a stronger twine to bind wreaths and make garlands.

GLASS JARS
All sizes. I insert them into containers to make the containers watertight. I like to use recycled yogurt jars for this.

COPPER WIRE
I love the sheen of it, and it can be recycled.

VASES
I prefer low bowls and vases with necks that are inverted (then I don't need tools to bind the stems). I love old pudding bowls, tureens, painted drinking glasses, and modern ceramics. If using bowls, just make sure they are deep enough to hold the stems in place. Pitchers also have a great shape for flower arrangements. For vases, I often use what I have around, small cups and bowls and mugs. But I also love to shop from local ceramic artists, thrift shops, antique stores, and terra-cotta producers.

SPRAY BOTTLE
For misting foliage

HEAVY-DUTY CLIPPERS
For woody stems

BUCKETS

CLOCHES
Or large glass vases to place over flowers to encourage them to open up.

BASKETS
All shapes and sizes

NEWSPAPER
Helpful when foraging for delicate flowers: Wrap them in paper for protection and set them in a basket or bucket to carry them home.

GARDENING GLOVES
Honestly, I rarely use gloves because I love the tactile experience of touching the plants and the earth. But if you have any sensitivities, you'll need gloves—especially for thorny stems and "sticky" flowers like daffodils and hyacinths and euphorbia.

STICKS
Sometimes I make little grids for the top of vases with sticks and twine for an all-natural flower frog.

A collection of ceramic and glass vases.

Vintage vases and traditional terra-cotta containers.

FLOWER-ARRANGING BASICS

It's helpful to know a few tricks of the trade when diving into flower arranging. Even when embracing the fragility of wildflowers and the different phases of a flower's life, of course we want to enjoy fresh cut flowers for as long as possible. With some very simple techniques, you can extend the vase life of your flowers, no chemicals needed! Timing the harvest, processing cut stems for maximum water intake, and proper care of your arrangement once home will ensure that most blooms will stay fresh for at least a week.

My flower "studio" is often an outdoor sink. You can make your own with a generous bucket and an outdoor table on which to keep scissors, vases, and arranging supplies.

Freshly foraged flowers, including thistles and roses.

Even in the conditioning bucket, you can start playing with different flower heights.

THE BASICS

For all flowers, you must recut the stems before you place them in water (see some specific types of cuts listed by variety on pages 81–83). Recutting the stems ensures the flowers will continue drinking water and stay hydrated. When cutting flowers from a garden or from the wild, always harvest in the morning or the evening when the flowers are the most hydrated. One exception is when you are picking flowers for drying to use in cooking or in fragrances; you will want to pick them in the hot sun while the oils are at their peak. If you do find yourself foraging for flowers to arrange in the hot sun, make sure, once you are inside, to place the stems in water in a cool, dark area and follow the normal conditioning process (see page 80) once they have recovered.

The stage the flower is in can ensure longer-lasting arrangements as well. As a general rule, harvest flowers when you see good color, buds that are halfway open, and leaves that are firm and hydrated. For many flowers, the blossoms may be open, but if you see fresh pollen on the petals or falling from the bloom, that means the flower is past its peak.

The main goals when processing flowers and placing them into a vase (and during their vase life) are to keep bacteria at bay and to keep them well hydrated. Bacteria gathering in the bottom of the vase will speed up the death process for flowers. That is why many people use flower food, a water-soluble powder that discourages bacteria growth. If you want to skip flower food, make sure to remove all the leaves that will be at or below the water level, keep your clippers sharp for a clean cut, and keep both vases and clippers pristine.

HOW TO REVIVE WILTING FLOWERS

A sink or tub full of water can help you revive a wilting arrangement. Submerge the entire flower in hot or lukewarm water for a few minutes and then recut the stems, place the flower in a fresh vase of water, and let sit overnight in a cool, dark area. This works especially well for roses and hydrangeas, but does not work as well for most white flowers or small, delicate blossoms.

Wilted roses that can be revived with a few minutes of care.

ECO-FRIENDLY TIPS

Here are a few tips for keeping your flower arrangements eco-friendly:

- Don't throw out water from your vases; use it to water your plants instead.
- Take care of the tools you have and try to buy used ones when possible.
- Compost leaves, stems, flower scraps, and dead flowers.
- Try to buy local and organic whenever possible.
- Use small amounts of water in the bottom of buckets; too much water can encourage bacteria and is a waste.
- Try not to buy flowers in plastic flower sleeves; ask your local suppliers if it is possible for them to wrap their flowers in paper and stop using plastic wrappers.
- Stay away from floral foam and try to use natural methods to compose arrangements.

HOW TO CONDITION FLOWERS AND WILDFLOWERS

Conditioning is the process of cleaning and trimming the flower stems so that they will last as long as possible in the vase. If you ever work at a flower shop, this will be one of your first jobs! When you bring flowers inside from the market, grocery store, farm, or garden, place them immediately into a bucket of water (remove any store packaging and recycle it). Place flowers grown from bulbs (tulips, hyacinths, daffodils, and so on) in cold water; room temperature water is fine for all other flowers.

NOTE: Daffodils exude a substance that kills other flowers, so make a clean cut on the stems and let them condition alone, in very cold water. You can combine them with other flowers later, but they may still affect the lifespan of the arrangement.

Let sit for a few hours if possible, then process the stems by removing all the leaves on the portion of the stem that is at or below the waterline. Next, recut the stem per the directions opposite. Place in a fresh bucket of water and let rest for a few more hours before beginning to arrange. This ensures that the flowers are as hydrated as possible before you begin to work with them. If you are concerned about freshness, look for green leaves and flowers just beginning to open.

If you are just on a walk and want to collect flowers, I suggest a slightly different process for gathering and conditioning. I gather the flowers in my hand, arranging them as I walk and adding to the bouquet as I go. When I get home, I bring the flowers inside and immediately put them into a vase of water. After a few hours, I recut the stems at an angle and return them to the water, letting the shortest stems touch the rim of the vase and making sure the weight and proportions work. Because wildflowers are very fragile, it is best not to touch them a lot. It's also best to harvest them in the morning or the evening when they are most hydrated. Realistically, though, if I am just foraging for home, it's all quite spontaneous.

HOW TO CUT FLOWER STEMS

As a general rule, cut all stems at a forty-five-degree angle with clean, sharp clippers. This allows the stem to take in as much water as possible. However, some types of stems and certain flower varieties need more specialized treatment.

CUSTOM CUTS

HOLLOW STEMS *SUCH AS* **DELPHINIUMS** AND **AMARYLLIS**

Place a wet cotton ball inside the stem to help it absorb water.

WOODY STEMS *SUCH AS* **LILACS, HYDRANGEAS**, AND **FRUIT BRANCHES**

Cut at an angle and then cut up the stem twice to create a cross shape at the bottom.

DAHLIAS

Dip the tip of the stem in boiling water before conditioning as usual.

POPPIES AND **EUPHORBIA**

These flowers produce sticky white liquid and need to have their stems sealed by holding the stem tip in a candle or match flame.

ROSES

Cut rose stems with care and with a very sharp tool: Bacteria thrive on bruised stems and will impede a long vase life.

The life of a fresh garden rose can be extended with the proper cut.

CUTTING WOODY STEMS

Cut at an angle with clippers or a sharp knife.

Cut into the stem toward the top, making the cut about a quarter inch (6mm).

Make a second cut into the stem to create a cross shape as shown.

CUTTING AN ANGLE

Hold one stem at a time, and position the clippers at a forty-five-degree angle.

Cut cleanly through the stem.

This angled cut exposes a good amount of surface area for water absorption; it's best for most stems, including the gomphrena stem shown here.

DETHORNING ROSES

Garden roses have a unique stem shape and thorn pattern.

Roses purchased from the market often have straight stems and very sharp, very large thorns—handle with care.

With a gloved hand, remove thorns by sliding a rag down the stem.

Instead of the method shown, you can also use a thorn stripper or sharp clippers to snip thorns individually, but be careful.

CONDITIONING DAHLIAS

Gather the stems loosely together.

Pour boiling water into a tempered glass vessel.

Cut the stem, then dip it in boiling water before conditioning as usual.

HOW TO ENCOURAGE FLOWERS TO OPEN FASTER

The secret to coaxing buds into bloom is gentle heat and humidity. Here are three time-tested approaches. Whichever you choose, always keep your flowers away from both heaters and drafts. You can place a large glass vase or cloche over the blossoms to create a sort of greenhouse. The warm temperature and humidity created by the cloche will encourage flowers to open. On the same principle, but without the glass cloche, try relocating the flowers you want to open to a warm (70°F/21°C) spot or room. Saving the simplest for last, experiment with placing your arrangement in warm water rather than cold or room temperature.

WHEN TO HARVEST FLOWERS

ANNUALS *SUCH AS*
COSMOS, MARIGOLDS, CLEOME
Flowers just barely open with no fresh pollen.

DAFFODILS
In bud.

HYACINTHS
Flowers open on the bottom of the stems.

HYDRANGEAS
The small flowers still very firm. If they feel soft, they will not last in the vase.

ROSES
Still in bud but beginning to open.

SPIKED FLOWERS *SUCH AS*
DELPHINIUM, LARKSPUR, FOXGLOVE, ETC.
On a tall stem with many small flowers; the flowers on the bottom are just opening.

SWEET PEAS
Open on the bottom of the stem and beginning to open on top.

TULIPS
Halfway open.

Honeysuckle, harvested while still in bud, with a mix of just-opened wildflowers.

BASICS OF DRIED FLOWERS

Drying flowers fell out of fashion for a while, but the preservation method is starting to become popular again. Preserving summer blooms this way for winter days is not unlike canning your summer produce. What a lovely way to hold on to your garden roses or herbs all year long, whether for decorative purposes or concocting them into recipes or herbal treatments. If you are drying flowers or plants to use for fragrance, or for cooking or medicinal use, harvest them when they are in the hot sun and their oil content is at its peak. These are a few methods I like to use for drying flowers.

AIR-DRYING

To air-dry fresh flowers, tie with twine and hang them in bunches of ten to twenty stems upside down in a dark, dry place—a barn or a dark shed is ideal. You can hang them on a laundry rack, or anywhere where they can have air circulation around the bunch. Keep the flowers away from sunlight, as it will fade them, and leave the stems to dry for about two weeks to a month, depending on temperature. If you are collecting seeds or blossoms (as you would from an elderflower, for example), you can put a paper bag around the flowers so that the small petals or seeds are collected.

WATER-DRYING

Water-drying works well for flowers like hydrangea (and has the added benefit of helping the flower retain its shape). This kind of drying often happens naturally if you forget to refill a vase with water. To water-dry, just put a small amount of water in the bottom of a vase of flowers and let it stay. The arrangement will eventually take up all the water and then begin to slowly dry. This usually takes about a week or two. You can also use this for other flowers that are firm to the touch, even when picked fresh. Try using with strawflower, yarrow, gomphrena, hydrangea, and seed heads like dill or fennel. It won't work well for flowers like roses that tend to bend their heads when wilting.

DRYING ON A SCREEN

Take the petals (or leaves) off the plant you want to dry and set them on a screen, then leave them in a dark, dry place for two weeks to a month. You can also dry petals and leaves in the sun, but don't leave them out too long or their color will fade. Either way, the screen-drying method is ideal because the petals will have a lot of air circulation and they can thoroughly dry. It's ideal when you are drying petals for potpourri or herbs for soaps or medicinal uses because you can be sure each petal or leaf is completely dry.

Garden roses hang drying.

Dried Limonium, solidago, gomphrena, and veronica.

HOW TO CARE FOR YOUR FLOWER ARRANGEMENT

Keep fresh flower arrangements away from bright sun and drafts. You may also want to keep your flowers away from fruit and vegetables since they exude an ethylene gas that speeds the death process of flowers. Remove dead blossoms and replace with fresh stems if you want. Most importantly, refresh the water in the vase every few days. You don't need to remove the arrangement to do this—simply pour the old water out while holding the arrangement to one side. Conserve water by pouring it directly into a plant that needs watering. Then, take the vase and just hold it under the faucet to add fresh water. You can fill the vase half full as long the water covers all the ends of the stems so they have access to the water.

A hydrangea-based arrangement is kept fresh with water and placement away from direct sunlight.

Roses are prepped for an arrangement.

Quinta da Ribafria, an abandoned villa in Sintra, where I forage and get inspired.

A Foraging Walk, Step-by-Step

FORAGING WALK

Let's go step-by-step on a foraging walk, taking note of the conditioning, composition, and even the floral meditation process that happens on these walks.

I carry small vases with me to give foraged stems some water—or sometimes I just put them in a vase as soon as I arrive home. Wildflowers don't like to be touched a lot, so I leave them intact unless I am foraging for a particular material or using them in a specific way.

I will leave these in the vase of water and then recut their stems the next morning, after I bring them home to rest in a cool, dark corner.

Wildflowers are fragile, so pick some extra stems since not all will survive the trip home. I am often on foot while foraging, so I normally use a gathering basket unless I am picking a delicate flower like poppies. (If you are picking poppies, bring some newspaper to wrap them in and a small bucket with a bit of water at the bottom.)

If you are foraging with a car, carry buckets of water—secured against spillage. When I am foraging for specific plants for a project, I forage them individually and sort them as I collect.

A hand-held bouquet of wild fennel, grasses, and dried Queen Anne's lace.

I start out on this livestock trail near the village of Sintra. This is where I begin my walk, looking for seasonal blooms, textures, and colors.

At the end of the livestock trail in Sintra, I discover a large patch of wild fennel at the end of the trail. At this time in August, most of the floral color has dried or faded, so it is a pleasant surprise to find the fresh fennel blossoms.

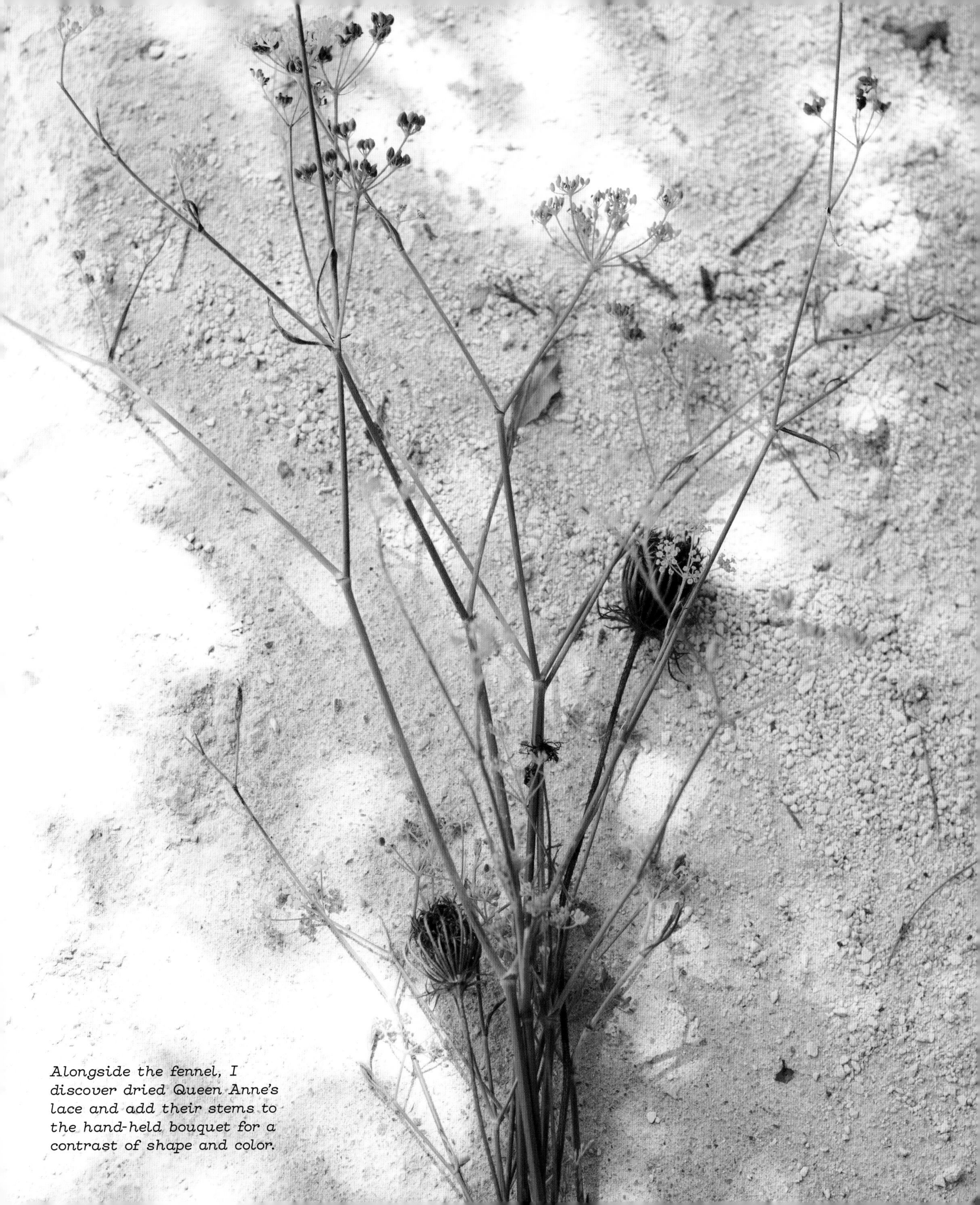

Alongside the fennel, I discover dried Queen Anne's lace and add their stems to the hand-held bouquet for a contrast of shape and color.

Next, I make my way to Quinta da Ribafria, an abandoned villa that is now a public park in Sintra with a forest, an orange grove, and trails.

At Quinta da Ribafria, I discover a stray petunia, milkwort and geranium. I am drawn to the vibrancy of the colors, which feel like jewels on my flower treasure hunt!

I arrange all the flowers with
the stems at different heights.

I add the flowers to the vase I've been carrying and fill it with water from the faucet at the villa. I include the colorful geranium and Petunia.

Later, I remove the geranium from the arrangement, succumbing to the soft yellows, greens, and browns with the purple. One arrangement is not better than the other, they are just different. The contrasting colors are vibrant, but the softness of the fennel and petunia with only the pop of the purple, feels more calming.

Hydrangea and jasmine growing in a home garden.

COMPOSITION

Now that you know how to harvest, care for, and even revive cut flowers, let's move to the basics of composing a flower arrangement. There are a few tricks of the trade that are helpful to know, even when embracing a "wild is better" mind-set with flowers. In fact, creating an effortless look can sometimes take a bit of effort, particularly if you are working with store-bought flowers.

An orange trumpet flower peeking up among morning glory vines.

A composition of summer wildflowers reflects the abundance of a summer meadow.

HOW TO HOLD STEMS IN PLACE

It's important to pay attention to the mechanics of the arrangement so that your stems aren't falling out of the vase! Here are some of the methods I use to hold stems in a vase or bouquet.

CONTAINER SHAPE

Choose a vase with an inverted top, or use a pitcher that is slightly inverted. The shape will hold the stems in place naturally without any need for additional tools.

HYDRANGEA SUPPORTS

Use a hydrangea as a base for an arrangement by poking the other flower stems through the hydrangea blossom (see the detailed how-to opposite). Hydrangeas are long-lasting and often dry right in the vase, so they make a perfect foundation.

PIN HOLDERS

Create a "flower pin holder" by twisting a separate bunch of stems into the bottom of a vase to hold the flower stems in place. This is inspired by the traditional metal pin holders that hold the stems perfectly when you push each one onto a pin; this natural version assists in holding stems in position, but not as securely or stiffly as the pin holder.

STONES AND ROCKS

Strategically places stones and rocks in the bottom of a vase to support the stems.

TWINE

Tie the stems together. You can use this method not just for bouquets, but also for making vase arrangements. Arrange the flowers in your hand (if you use the right vase you can then place them right into the vase, adjusting stem heights as needed), tie with twine, then place into the vase so that the lowest flowers sit at the rim of the vase. You can leave it at that, but if you feel the arrangement needs more, now that it is stable, you can then begin to add stems to the bouquet that is bound with twine.

HOW TO USE HYDRANGEA SUPPORTS

1

Gather the flowers that you will use with the hydrangea to get a sense of the overall arrangement.

2

Cut the stems of the hydrangeas short so the flowers rest on the rim of the vase. Use two or three and cross their stems for stability.

3

Add sturdy blooms, like dahlias and sedum, to create a structure. Place each stem into the hydrangea blooms, so the structure of the hydrangeas helps them stand upright.

4

Add some grasses and roses in all phases of bloom for added textural interest.

5

The finished arrangement should be balanced, yet casual. Notice how the hydrangea blends in and works as a beautiful base and support.

HOW TO EDIT A STEM

It can be important to edit stems both for visual appeal and to make the mechanics of the arrangement secure; it also makes it easier for the flowers to drink water, so they will last longer. See the images below: If there are multiple stems above a node, I cut them off and separate them. This creates several stems instead of one. You can pair the small buds with the longer stems in the arrangement to get the same look the flower had before, but editing the stem makes the technical process of arranging much easier because it gives you more control over the stems.

Examine the stem for multiple offshoots that can each become individual elements: Separate these from the main stem, as shown.

CREATING AN ARRANGEMENT

Composition can make all the difference with just a few stems. On the following spread you see a small arrangement I've made with garlic flowers. It becomes something simple and elegant when the plants are placed in thoughtful ways instead of all at the same height. Each one is at a different height and the delicate grasses follow the lines of the focal flowers (the garlic flowers). I always like to have a flower touching the rim of the vase. The greens I've used here are chartreuse to contrast the lavender flowers.

TIPS FOR COMPOSING AN ARRANGEMENT

VARY HEIGHT

Add flowers at differing heights. Leave some flowers shorter to cover the rim of the vase and others tall to draw the eye upward, and to all heights in between!

GROUP SMALL BLOOMS

Add small, delicate flowers in clumps for stability and visual focus.

CREATE NEGATIVE SPACE

Add tall, lightweight materials, such as meadow grasses, dill stems, or ranunculus buds., that add surprising proportion but won't disturb the physical balance of your arrangement.

CONSIDER SCALE

Think about the size of your arrangement and choose the container accordingly. The traditional proportion is one-third vase to two-thirds flower arrangement, but I like to mix this up by making the flowers just a bit higher or much lower, nearly to half and half, with some lighter grasses for additional height.

USE COLOR

Embrace one floral variety or a single color; sometimes the most visual impact comes through variations of one hue.

ADD TEXTURE

Use a variety of plant materials.

EMBRACE REPETITION

Using multiples of the same blossoms creates a place for your eye to focus.

THE ELEMENTS OF A FLORAL ARRANGEMENT

FOCAL FLOWERS
Like the peony here, they grab your attention, your eye goes right to them.

FILLER FLOWERS
They usually have lots of small blossoms and fill space.

SECONDARY FLOWERS
Less showy than focal flowers, they often add texture.

GREENERY
Offers color contrast and can add stability to the arrangement.

A SIMPLE LESSON IN COLOR AND COMPOSITION

This is an arrangement I made with garlic flowers. The flowers were picked in the evening from a local parking lot. Lavender and chartreuse are complementary colors (for more ideas about color, see the next section on color theory) and the blue of the vase is a soft companion color to the lavender garlic blossoms. There is contrast but also subtlety, which extends to the neutral grasses.

COMPOSITION STEP-BY-STEP

1

For this composition, I used garlic flowers, Brazilian pepperberry greens, wild grass, and sticky weed greens. Please note that this plant is allergenic; use gloves if you have sensitive skin.

2

Position the tallest stem first. Here, it is the garlic flower stem.

3

Next, add some shorter stems, cut so that their blossoms almost touch the rim of the vase.

4

Add some greenery to fill out the composition.

5

Add the grasses last, for texture and to create some balance.

COLOR THEORY FOR FLOWERS

If you feel lost when it comes to color, consulting the color wheel and even learning a bit of the basics can help you get started. I tend toward monochromatic and analogous color schemes, which are close to each other on the color wheel. Any time I use complementary colors (based across from each other on the color wheel), I balance the stronger colors with something pale, and I don't use many stems in the dominant colors. I also like playing with variations of saturation for a single color. For example, the aster and geranium arrangement on page 140 relies on hues in shades of pink and burgundy.

A complementary color scheme of yellow and purple creates a vibrant little posy from a homegrown garden.

GO-TO COLOR COMBOS FOR FLORAL ARRANGEMENTS

ANALOGOUS
Two to three colors next to each other on the color wheel.

MONOCHROMATIC
Varying saturations of a single color.

COMPLEMENTARY
Two colors across from each other on the color wheel (such as red and green).

TRIADIC
Three colors evenly spaced on the color wheel (such as orange, purple, and green).

Ingredients in colors all next to each other on the color wheel (purple, pink, red).

part II

FLOWERS by ROOM

Jardines de bolsillo

chapter 1 / *Living Room*

The living room—often where we entertain guests, spend restful afternoons on the sofa, or put in productive days at a corner desk—can be infused with floral elements to bring vitality to our days. Perhaps you take a break midday to search for wild roses on a trail and later add them to your workspace; or spend an afternoon making party arrangements for a spontaneous gathering. Flowers can play a central role in this busy, multifunctional space.

I've often pieced together thrift store finds and my grandmother's hand-me-downs to decorate my homes, but it's the finishing touches—a tree branch or a bouquet of wild roses—that have always made my spaces feel full of life, dynamic, and even on-trend. Adding foraged wildflowers or cutting garden blossoms can completely refresh your living space without costing extra money.

POPPY WELCOME

This meadow-like arrangement of wispy poppies, daisies, and nasturtium vines invokes spring tradition and positive energy. Tucked into a bright, vintage bowl, it's perfect for greeting visitors at the door.

Each May in Portugal we celebrate *Dia da Espiga* (which literally translates to "day of the ear of grain"), a holiday with roots in both Catholicism and ancient pagan springtime rituals. Forty days after Easter, flower sellers flood the streets in Lisbon, selling foraged bouquets of poppies, daisies, olive branches, wheat, rosemary, wild grasses, and vines. Those of us in the countryside gather the bouquets ourselves. Although many of the rituals of religion have faded, this sweet tradition lives on.

Festivity and symbolism are the main selling points for these wild sprays: Olive branches are used to represent peace, poppies for love, rosemary for health, wheat for prosperity, daisies for gold or good fortune, and vines for joy. The bouquets are not hardy—and in fact they often wilt quickly in the hot Lisbon sun—but many people keep them to dry inside the house, and then hang them by a door to bring in luck and good energy all year.

The varieties themselves can be loosely interpreted. For instance, the daisies might be chamomile, garland daisies, Spanish daisies, or feverfew. You can use what you have nearby, just as the flower sellers do. I especially love the concept of vines symbolizing joy and used that as my inspiration for this arrangement.

To secure the blooms and grasses, we are creating our own natural version of a stem holder, which will make the entire composition feel much less stiff than if we used a flower frog. Wrapping vines around each other can create a reinforcement in the bottom of the container to hold delicate stems in place. I wouldn't travel with this arrangement, but it's stable enough to sit in your house—no need to use synthetic tools. The poppies will last about three days, the other flowers up to a week, if the bowl is refilled with fresh water.

You will need

- 10 poppies
- A selection of other wildflowers, vines, and grasses—the arrangement shown contains feverfew and wild daisies (2–3 stems each), nasturtium vines with flowers (3), and meadow grasses (about 10 stems)
- 5 pliable but strong vines, such as ivy, for the flower frog
- A small bowl, at least 4 inches (10 cm) tall

Dia da Espiga bouquets at a flower seller's stall in Lisbon.

1. Gather the ingredients for your arrangement.

2. Trim and condition the flowers and vines as soon as possible after bringing them home from foraging or shopping. The poppies will need a special treatment before you begin working with them. (See page 81 for details on conditioning poppies.)

3. Bundle the vines in the bottom of the bowl to create the flower frog, which will provide structure to the arrangement. Next, add some poppies that still have their greenery. This will help build a shape.

4. Fill the gaps using the feverfew, daisies, and grasses in varying lengths to create the "meadow" effect, and then tuck in the nasturtium vines, letting them fall over the bowl.

5. Finish by adding in a few more wispy poppies.

A SUIVRE)19
NUMEROS
108. 109. 110. 111. 112. 113
1987 JANVIER / JUIN

WILD ROSES FOR THE DESK

Filling a pitcher with freshly picked blossoms and greens is a quick and easy way to bring summer inside and brighten up even a pragmatic corner of your living room or office.

Tossing stems of summer wildflowers into a pitcher with minimal adjustments feels like an essential summer ritual. I've collected all sorts of vintage pitchers over the years to do just that. Their narrow necks provide easy structure for free-flowing arrangements without the need for any sort of tying or mechanics.

For this meadow-on-your-desk, go for a walk and gather stems of wildflowers in a lot of different varieties. Look beyond floral elements and see if you can find interesting textural combinations—different leaf patterns and scales, contrasting greens, and dried grasses will offer color contrast, even in a non-floral arrangement.

Here I used cow parsley, thistles, wild roses, willowherb, campion, grasses, and even dandelions. Notice that as the campion dried and fell from the vase, I left it. I don't like flowers to look too perfect, as though they are fake. In nature we see wilting petals, dried bits, and less than pristine blooms, and they all create interesting textures within an arrangement. Embrace this imperfection and you can create effortless fresh flower displays.

You will need

- 5 to 7 bunches of meadow grasses
- A mix of foraged wildflowers—whatever you find on your walk
- A tall pitcher with a narrow neck

Wild roses make a beautiful arrangement, but gather with care; they can't tolerate lots of handling.

1. Gather a basket of wildflowers, arranging them in your hand as you go, handling them as minimally as possible.

2. Aim for a mix of taller stems, such as grasses and wild roses, along with shorter ones to create space in the arrangement; include a few unruly vines for movement.

3. Cut the stems when you bring them home and place them immediately into a pitcher with fresh water. After a few hours, or overnight, refresh the water (use the old water to water plants). Remove the leaves from the wildflowers, trim the stem ends—less than an inch (2.5 cm)—and place them back in the vessel. This will ensure that the flowers are hydrated properly, but with minimal handling.

BOOKSHELF ROSES WITH HERBS

An elegant, low-profile arrangement of roses and herbs slips right into a bookshelf vignette.

Herbs aren't often seen outside of the kitchen or bath, but they make for a pleasant surprise that most people can't help but sniff and touch! This rose arrangement—enhanced by long stems of thyme, hyssop and jasmine, and grounded by seasonal greens—enlivens an otherwise subdued shelf.

I composed this entirely from farmers' market ingredients: culinary herbs, garden rose bunches from an edible flower seller, and foraged greens that were originally part of a mixed bouquet that I deconstructed. You could easily do the same with a dozen small sweetheart roses from the grocery store and herbs from the veggie aisle or clipped from pots.

Keep in mind that a fragile or non-watertight vessel, like this handmade, V-shaped ceramic bowl, will need an inner liner to hold the arrangement and its water. Here, a small glass jar does the trick.

You will need

- 10 stems of roses, and several bunches of herbs and greens, in any combination
- The bouquet shown includes garden roses, jasmine vines, eucalyptus, hyssop, thyme, and pokeweed.
- A low bowl or vase, about 8 inches (20 cm) wide, at least 4 inches (10 cm) tall
- A small glass jar, such as a mason jar, that fits inside the vessel

A bookshelf comes to life with the addition of roses, herbs, and foraged pokeweed.

1. Gather roses, herbs, and greens in an appealing combination from an edible flower seller at the farmers' market or grocery store. Once home, trim the stems at an angle and set them in water for a few hours. The roses may need their thorns removed.

2. Next, place the mason jar (or other glass container) inside the ceramic bowl to make a watertight container for the flowers.

3. Create a shape for the arrangement using the herbs, greens, and garden roses by placing their stems in the small glass container.

4. Begin adding stems in proportion with the vessel. Add the long jasmine stems and pokeweed after the structure of the arrangement is in place, using their long stems to create more negative space. Tuck the hyssop into the bouquet in bunches and let the longer stems extend beyond the lip of the ceramic bowl.

MEADOW OFFERING

A tied bouquet of gomphrena, grasses, and dandelions, delivered in a terra-cotta cup instead of paper wrapping, makes a sweet hostess gift.

This useful and beautiful little bouquet is inspired by summer meadows and is a thoughtful gesture for a friend. Gomphrena, with its cheerful, clover-like blooms, is a popular crop in Portugal; it's quite common to see it both fresh and dried, as it's grown organically for use in teas—in fact, your bouquet recipient can dry the gomphrena and later use it for tea herself as long as you know where and how the stems have been grown. If that is your intention, you will want to purchase directly from an organic farmer. I bought these at the flower market, then paired them with foraged grass, sedum for filler, false dandelions (leafy plants that look very much like dandelions), and greens.

The bouquet is simply tied so you can transport it without water, then recut the stems, transfer it to a little vase or drinking glass, and add water when you arrive (leaving your friend the terra-cotta cup to keep, of course). The tied bouquet is a basic flower-arranging technique that can be applied to large-scale or small-scale bouquets, and it works especially well with any sort of filler flowers. I love to make all-filler bouquets, particularly if I don't have access to a lot of foraged materials, as they will offer a similar look—especially if you can pop in a few weeds or grasses taken from the sidewalk or a local park. If all you have are grocery-store flowers and some dandelions and grasses, use them!

You will need

- 10 stems of gomphrena
- 5 stems of meadow grasses
- A mix of flowering weeds or wildflowers: Used here are 3 stems each of false dandelion and sedum; you could also use elderflower or Queen Anne's lace.
- Florist's twine
- Terra-cotta cup (or small jar), plus a tag and ribbon, if desired, for presentation

Locally grown gomphrena conditioning in buckets.

1. Gather the flowers and filler—such as these sedum, foraged grasses, false dandelion, and gomphrena—from your flower market and/or local foraging spots.
2. Once home, remove the leaves from the bottom half of all of the stems. Cut an 8-inch (20-cm) length of twine and set it on the work surface.
3. Start by holding a few stems in one hand and begin adding flowers with the other hand. Each time you add a stem, turn the bouquet—this will ensure you create a bouquet that looks balanced and pretty from all sides.

4. Keep the gomphrena taller in some places than other stems and try to add them at different heights. Having both taller and shorter elements creates visual interest.
5. When all stems are added and you're happy with the shape, bind the bouquet by wrapping the twine around the stems a few times and then tying it in a knot.
6. Cut the stems so that the lowest flowers are resting on the rim of the cup or jar. Finish with a tag and ribbon, if you like.

A party table is prepped with ceramics, flower petals, and the hand-braided hanging wreath.

STRAWFLOWER HANGING WREATH

Braid fresh flowers into a hanging wreath for a party—then let it dry and leave it up all year!

A hanging wreath adds a sense of playfulness to a room and is, of course, perfect for a little party or special occasion. This fresh-to-dry project can be made with just-picked flowers, but it's also a great way to use up ones that are on their way out—blooms that are wilting but still have pliable stems can be braided and will dry nicely.

Bay leaves, gomphrena, and strawflowers are especially good choices for an arrangement like this one, as they're all very long-lasting and dry well. The colors of strawflowers feel almost unreal (in the best way!) and bring a sense of whimsy to the project. This wreath is light enough that it can easily be suspended from a ceiling hook, in part because it's made without wire. Stems of millet and statice give it stability, and the braided strawflower stems keep the blossoms in place. These braiding and weaving methods can be applied to wreaths of all sorts. A hanging flower wreath is a focal point, so you won't need other large arrangements or elaborate styling. Here, I hung the wreath over a table covered in pink linen and added modern ceramics, tiny arrangements, and a few scattered flower petals.

You will need

- 8 stems of small bay leaves
- 25 stems of strawflower
- 3 to 5 stems of common millet or pampas grass
- 5 stems of statice
- 5 stems of gomphrena
- Florist's twine or string, for hanging

The finished
strawflower wreath!

1. Gather your ingredients at a wholesale flower or farmers' market. Once home, set them out in bunches on your work surface, and remove the leaves from the strawflower stems. If the flowers have been in water, be sure to remove any debris or excess leaves from the stems and let them dry out before working with them.

2. Create a base with the millet and statice, working with a few stems at a time.

3. Curve them into a wreath shape, spacing the blooms at intervals and twisting the stems around each other to secure.

4. Braid bunches of strawflower together, just as you would hair, alternating between three and five stems per bunch. Once braided, weave the bunches of strawflower into the wreath base, adding more statice for security if needed.

5. Fill in the wreath with bay leaves between each bunch, then weave in gomphrena between the stems.

6. Create a hanger by tying lengths of string or twine to opposite sides of the wreath. Display away from direct sunlight, if possible, or it will fade quickly.

GERANIUM AND ASTER TABLE BOUQUETS

With a little restyling and a few wild touches, you can transform ordinary grocery-store flowers into a blowsy, gardeny display for a last-minute party.

File this under *how to do grocery-store flowers right*. Though I sourced this arrangement of saturated pink asters and geraniums, with airy fillers of foraged salvia and gaura, from a variety of places, you could get the main ingredients at a supermarket, then style them with a few found elements. Use the technique here as a starting point.

To make the most of flowers from the store, you'll have to do some editing. With a mixed bouquet, start by taking out what isn't working: Perhaps keep the greens and fillers and don't use the focal flowers, or separate them into different containers. Stick with one color if you can—a monochromatic approach allows your eye to focus on the arrangement as a whole and its different textures. Add some wispy, wild-looking elements—foraged soft grasses from the patio or by the sidewalk, or cuttings from a plant in the garden—to provide movement and lightness. And finally, cut down stems that are too tall and stiff. By cutting some quite short, so that the flowers sit at the rim of the vase, you add more visual interest, negative space, and drama.

For an easy, elegant presentation, layer your arrangements in glasses of different heights and styles. Mix textured pieces, like the thrift-store find with gold etching here, alongside plain drinking glasses. The nice thing about glassware is that it is unobtrusive, letting you focus on the flowers themselves. It feels like you aren't trying too hard, in the best way.

You will need

- A mix of grocery-store flowers and fillers, cuttings, and foraged finds
- The bouquets shown here include bedding geranium, asters, gaura, salvia, sweet William, false dandelion, tarragon, carnation, and Virginia creeper vine.
- A selection of drinking glasses

Arrangements in vintage drinking glasses are ready for the table.

How-to

1. Bring store-bought bouquets home and separate the blooms, greens, and fillers. Remove leaves from the bottom of stems and let sit in water for a few hours before working with them. Be sure to recut the stems at an angle as you add them into the arrangements.

2. Start by placing asters of different heights in a glass.

3. Add in the sweet William, the geranium, and carnations. Keep larger blossoms near the rim of the vase.

4. Add gaura, salvia, tarragon, and false dandelion. Since these are more delicate, add them in groups to balance the arrangement.

5. Add in more carnations, asters, and geranium as needed. Finish off with the Virginia creeper vines at a taller height to add movement and negative space to the arrangement.

6. Repeat with more glasses, flowers, and fillers. Display around the living room or down the center of a family-style dinner table. Give the bouquets to your guests at the end of the evening!

Jardines de bolsillo
Proyectos japoneses
contemporáneos en miniatura

DANDELION STUDY

Even if all you have are some leggy dandelions growing along the sidewalk and a small drinking glass, you've got an artful display in the making.

Yes, weeds can be an art piece in your house! Dress a mantel with some thoughtfully arranged grasses and dandelions. This is a minimalist look—with just a few of each element, cut at varying heights, it's all about the restful, negative space. There is not a lot of structure to this arrangement: A few stems lean on one side of the glass and a few on the other. But the simple composition makes the weeds feel special. I think you'll find this airy arrangement to be refined and surprising in the best way. Think of it as a floral meditation. Take a few moments in your day to recognize the intrigue of a common weed, taking it inside to appreciate and to integrate a bit of nature into a corner of your home.

You will need

- 2 stems of dandelion or false dandelion
- 3 stems of grasses
- Drinking glass

How-to

1. Gather the stems from along the sidewalk or another foraging spot and trim them to the desired lengths.

2. Place the stems in the glass one at a time, letting the tall stem lean left, and the short stem lean right, then add grass to crisscross the stems so they balance each other.

DRIED HIBISCUS PETAL GARLAND

Hibiscus petals aren't just delicious in teas and cooking; they also have lovely texture and fragrance when used as a dried display. String them with other seasonal ingredients such as olive leaves and globe amaranth, and you can enjoy them all winter.

A handmade garland adds a *je ne sais quoi* element to a living room when hung over a bookshelf or sideboard. This one is made with hibiscus petals, which dry beautifully, add intense texture to bowls of other dried ingredients, and are even prettier when strung together.

When hanging a garland, think about creating a small vignette: Displaying items like this—using one new thing to set off other things you already own—is something I've been doing forever and it's such a beautiful way to appreciate what you have and to freshen up a room.

Look for petals, leaves, and seed pods at specialty tea shops, or collect them yourself from foraging spots and set them on a screen to dry. Then string them—I love doing this while watching a movie or listening to music.

You will need

- A variety of dried flowers and leaves in half-ounce (15-g) quantities. Look for different colors and scales of leaves, pods, and flowers. The garland shown includes hibiscus, olive, lavender, globe amaranth, and cassia.
- Sewing needle
- Good quality, all-purpose sewing thread (cotton)

How-to

1. Gather a selection of dried petals, leaves, and seed pods from a tea shop (or collect and air dry them yourself). Cut a 48-inch (1.22-m) length of thread, double it for extra strength, and thread the needle.

2. Push the needle through the flowers one at a time, finding the strongest part of each one to pierce. To create interesting texture, vary the angle for each petal—sometimes pushing the needle through the middle, sometimes from top to bottom—and weave the thread in and out for the olive leaves. Group some hibiscus petals and cassia together and alternate the patterns as you go. Don't worry if the string tangles a bit or if the pattern isn't consistent—it adds to the charm!

3. Tie off the end of the thread, making a 1-inch (2.5-cm) loop, and hang in a corner, on a mantel, over a window, or on a shelf.

chapter 2 /

Kitchen

The kitchen always feels a little bit more under control after a trip to the farmers' market, when I have flowers, herbs, and a bowl of freshly picked produce right at hand. The sellers at local markets in Portugal regularly add in a free bunch of parsley or cilantro with my purchases, and I'm often tempted to bring home an old-world garlic braid or dried oregano. I hang them up in the kitchen alongside the bowls full of fresh produce, and I'm ready for the week.

Whether you buy swags of herbs or make your own, toss petals onto freshly made meals, or create a tea wreath, there is always plenty of room for flowers and herbs in the kitchen. You can purchase edible flowers from the farmers' market or grow them at home—then use them to infuse your foods with their subtle, and sometimes spicy, flavors. Of course, they look beautiful, too.

An edible bouquet of nasturtium and lavender site on a kitchen shelf.

BASICS OF EDIBLE FLOWERS

Never eat a flower that you can't identify with 100 percent surety. Note that for some flowers, only certain parts can be eaten. For instance, elderflower stems and leaves are not edible; the flowers are not edible raw, but the petals can be used to infuse syrups and liquids. A lot of misinformation lives on the Internet when it comes to edible flowers. Even a bloom used to decorate a cake might not necessarily be edible, so assume nothing and always do your research. Also note that some flowers and plants are edible but can be difficult to digest or have an unpleasant taste—chicory is one example (see Safety, page 238).

FORAGING AT THE MARKET

Always look for edible flowers that are unsprayed and organic. If you want to forage, get access to a local foraging expert and tag along with him or her to learn what you can find locally and safely eat.

To process purchased items you are unsure are clean, rinse or spray the edibles in three parts water to one part vinegar. For heftier flowers, you can soak them and then let them dry. It's important to not add too much vinegar or your flowers will have a strong vinegar taste. Remove any parts that cannot be consumed, such as the fuzzy stems of borage and elderflower stems and leaves. Marigolds should have the white part of the petals removed, as they taste bitter.

A FEW OF MY FAVORITE EDIBLE FLOWERS TO GROW OR PURCHASE

- Arugula flowers
- Bachelor's buttons
- Basil flowers
- Borage flowers
- Calendula
- Chive blossoms
- Dianthus
- Elderflower (petals only)
- Garlic flowers
- Jerusalem artichoke flowers
- Lilac
- Marigolds
- Nasturtium
- Pansies
- Rosemary flowers
- Roses
- Thyme flowers
- Violets

A botanical wreath keeps herbs close at hand in the kitchen.

SUMMER TISANE WREATH

Weaving your favorite floral and herb infusions into a wreath can provide easy access to cups of tea all year long.

Having a kitchen stocked up and prepared can go beyond bowls of fruit, pitchers of herbs, or garlic strands. This fresh-to-dry project—made with your favorite herbs for infusing—invites conversation, clipping, and many cups of tea to enjoy with friends. You can use fresh or wilted herbs for this project, as long as their stems are not too dry to be pliable.

The finished wreath will look charming in a breakfast nook, hanging in a corner of the kitchen, or over a table where people gather. Keep clippers and teacups nearby, along with a tea strainer to make hot or cold infusions. You can replenish the wreath with herbs as desired.

You will need

- Assorted herbs and flowers appropriate for tea, such as:
- Hyssop
- Lavender
- Lemon verbena
- Mint
- Nettle
- Rose
- Thyme
- Grape vine base (see page 155 for instructions to make this wreath)
- Twine or ribbon, for hanging

1. Gather herbs in the garden or at the farmers' market.
2. Once home, take leaves off the bottom portion of the stems of all the herbs (lavender is shown here) so that you can handle and weave them easily.
3. Twist together the vines to form the wreath base.
4. Starting with the lavender (or another sturdy herb) weave the herb stems between the branches of the wreath form.
5. Tuck fillers, like verbena, into open spots. Add the most delicate flowers or herbs last. Continue weaving until the wreath is at the desired fullness.
6. Finish off the wreath with a garden rose and hyssop filler.
7. Tie twine or ribbon to the base to hang the wreath. Use the wreath as a handy supply of kitchen herbs: Here, mint leaves are infused for tea.

BASIL FLOWER BUTTER

Herbal butters are nothing new, but butter infused with the flowers of the basil plant makes for a pleasant surprise when spread on a piece of fresh bread or toast.

Late in the season, when basil has flowered, its leaves often become too strong or bitter for many people's taste. Those flowers, however, offer a less intense—even sweet—flavor. As your basil flowers, pluck the pretty blooms and use them to add a soft, subtly spicy floral flavor to whipped butter. It will look wonderful on the table (garnished with more of the same flowers) and will be a delight as it melts in your mouth.

Bunches of basil in bloom.

You will need

- 1 bunch of freshly flowering basil
- 2 cups (4 sticks) of butter (salted or unsalted, according to your taste), softened
- 2 to 3 tablespoons of milk or cream
- Parchment paper

How-to

1. Check the farmers' market for a bunch of basil that is flowering, or pluck blooms from your own basil plants. Remove the basil flowers, collecting about ¼ cup. Clean off any debris or old petals. The basil flowers are too delicate for washing, so set them on a towel after removing the old petals and leaves to make sure there are no bugs or other unwanted critters. If you see any critters you can shake them off.
2. Combine the flowers with the softened butter and a touch of milk or cream in a bowl. Blend by hand until the desired consistency is reached and the flowers are distributed evenly.
3. Wrap the butter in parchment and let it sit in the fridge for a few hours to solidify again. To serve, add basil leaves and/or flowers for garnish.

QUICK PICKLED GARLIC FLOWERS

Add an unexpected kick to salads, sandwiches, bowls, and main dishes with pickled flowers.

Want an imaginative garnish? Normally when people think of cooking with flowers, they think of sweet dishes, so surprising your guests with an acidic little flower blossom can be fun! I like to use marigolds and garlic flowers for this, as I found that their sturdy petals work quite well. If you do use marigolds, cut off the white portion, as it has a bitter taste. You can also mix some flowers in with veggies when you are quick pickling.

A quick pickle recipe for edible flowers.

You will need

- 12 stems of quality edible flowers, such as garlic flowers or marigolds
- 1 cup (240 ml) white wine vinegar
- 1 teaspoon peppercorns
- Salt and pepper
- Fresh herb leaves (optional)
- Borage flowers, for garnish (optional)

How-to

1. Purchase edible flowers at a farmers' market or cut from your own plants. Prepare them by rinsing in cold water or a vinegar mixture (see page 151).

2. In a bowl, combine white wine vinegar, peppercorns, dashes of salt and pepper, and a splash of cold water. Mix in some snipped fresh herbs, if desired. Add the flowers and let them marinate a few hours, either on the counter or in the fridge. Drain before using.

3. Add to salads or open-faced sandwiches. Use them the same day you make them.

WARM PLUMS WITH LAVENDER-INFUSED MASCARPONE AND THYME BLOSSOMS

This dessert of vivid plums—on a base of cream and mascarpone gently infused with lavender and topped with flowering thyme—celebrates the fleeting beauty of summer's showstoppers.

Cooking according to color and season comes naturally to flower-lovers, I suppose. As I do when composing a bouquet, I tend to start planning a dish with a single color. I gather my ingredients from whatever seasonal palette catches my eye and edit as I go. For this luscious dessert, after spotting bundles of thyme, purple basil, lavender, and baskets of plums all in aubergine hues at my local market, I began experimenting. Eventually I edited out the basil and then delved into the other violet hues, adding a bit of decadent cream and mascarpone, blossoms of lavender and thyme, and in the end arrived at a lavender-infused summer dish that will have even the most apprehensive flower-eaters licking their lips!

In season, local produce always offers the richest flavors and colors, and freshest product. Plum season arrives midsummer in Portugal, first with the green plums—which I like to use in arrangements with their sprawling leafed branches—small purple plums a bit larger than cherries, and your standard juicy purple plums. In drier parts of the country, you'll find lavender growing wild, a variety we often call Spanish lavender, at the same time. High-quality honey (Portugal's honey production is immense and diverse, making it common to find anything from a lavender honey to a creamed orange flower honey at regular supermarkets) makes this plate feel extra rich.

Of course, you don't need to come to Portugal to make it. You can mix and match this recipe to suit your local harvests and resources. Consider rose petals paired with strawberries or raspberries. Once you start making floral-infused cream desserts, it's difficult to stop!

You will need

- 3 to 4 stems lavender blossoms
- A few stems thyme blossoms
- 1 cup (240 ml) heavy cream
- 1 dozen small to medium purple plums, pitted and halved
- 1 teaspoon butter
- ¼ cup (60 ml) mascarpone
- Honey to taste

Warm plums over a bed of lavender-infused mascarpone.

Lavender and plum ingredients waiting to be prepared.

1. Purchase blooming lavender and thyme at a farmers' market, or pick from your garden. At home, prepare them by removing the blossoms from the stems. Rinse quickly in cold water and then leave on a towel to dry.

2. In a saucepan, warm the cream a bit, then add the blossoms (reserving a few for garnish), and take off the heat as it starts to bubble. Let sit about 15 minutes and then strain the blossoms out. Let the cream cool.

3. Meanwhile, in a frying pan, barely warm the plums with butter. After the cream has cooled, stir it into the mascarpone and blend well with a spoon.

4. Plate the mascarpone, top with plums, and drizzle with honey to taste. Finish each dish with 2–3 thyme blossoms.

CULINARY HERB AND FLOWER SWAG

Even if you don't live in the countryside, you can bring a bit of an old-world, rustic energy to your kitchen by hanging up strands of garlic, onions, or herbs.

At my local market I am often inspired to pick up a strand of onions to bring life and abundance to my kitchen, never mind the practicality of having seasonings and supplies right at hand for the week.

Whether you buy herbs from a green market or grocery store, or grow them in your garden, you can bundle them and hang them in your kitchen to dry—and you don't even need to wait before you clip from the stems to add a needed bit of sage to a chicken or some thyme blossoms to a salad. In fact, having the ingredients just where they are needed might spark inspiration as you prepare your evening meals. Aside from the convenience and no-waste approach, swags add warmth, color, and fragrance to a kitchen in just a few minutes, making this a practical little project for anyone who is just trying to stay ahead of a busy week!

You will need

- 6 bunches of fresh herbs and edible flowers in any combination
- The swag pictured includes sage, thyme, lemon balm, and basil, and stems of lavender and scented geranium.
- Florist's twine, or any decorative fiber, for binding

A traditional swag of oregano flowers creates a charming kitchen vignette.

An herbal swag adds scent and welcome to an entryway.

1. Collect 6 bunches of herbs, including some with flowers, from a garden, market, or supermarket. Rinse and let dry completely (damp herbs will mold in the swag).

2. Trim the stems and leaves to your desired amount of tidiness and length. Next, cut a 6-inch (15-cm) length of twine and set aside in easy reach.

3. Lay the herbs and flowers on a table and gather them in your hand, placing shorter stems at the top and up front, longer ones in the back. Aim for a mix, and bunch stems together neatly but not to too tightly. The arrangement should be about 5 inches (12 cm) in diameter, similar in size to a handful of dried spaghetti.

4. Holding the flowers in your nondominant hand, bind the bunch with the twine. Knot the first loop of twine, to keep it in place, then wrap the stems from top to bottom as shown. Secure with a small knot and tuck in the tail.

5. Hang the swag away from too much heat and sunlight for best results. The herbs should become totally dry in about two weeks, but feel free to pinch or trim off small amounts to use for seasoning while the swag is drying.

6. Once the swag is completely dry, you can leave it hanging up or store the dried bundle in a jar—or several jars if you want to separate each herb. To use in cooking, pinch or cut off a small amount and reinvigorate the plant oils by rubbing the dried material between your palms before tossing them into the pot or dish.

FLOWER FRITTATA

With spicy nasturtium leaves and flowers of cucumber, Jerusalem artichoke, and calendula, this egg dish makes a stunning display. Make them individually for guests or a larger one for the center of the table.

One of the dishes I make for breakfast the most often, this is somewhere between an open-faced omelette and a frittata, with hints of German pancake. I am always looking for ways to incorporate flowers into savory dishes and alongside eggs, and this dish does it beautifully.

For the perfect, light texture, be sure to whip the eggs until they are frothy. I love what a difference this makes. I remember the first time I made it, I was so inspired by how a few basic ingredients can turn magical just by whipping them with a fork! You'll stir some of the blooms, along with chopped herbs, right into the egg mixture, and later pile others on top for a garnish, almost like a little cake! It's airy and spicy—great with a big cup of coffee in the morning. This version is for an individual serving for a dinner-size plate, but you could make a larger one in a cast-iron pan.

You will need

- 1 to 2 calendula blossoms
- Around 5 nasturtium leaves and flowers
- 5 cucumber flowers (optional)
- 1 to 2 Jerusalem artichoke flowers (optional)
- Chopped herbs such as dill, basil, or parsley
- 2 to 3 eggs
- A splash of milk or cream (optional)
- Salt and pepper
- 1 tablespoon butter
- Plain Greek yogurt, to taste, about 4 tablespoons (60 ml)
- Chili oil, to taste

Unlike many cooking ingredients, these recipe leftovers [nasturtium] also make a pretty decoration.

1. Purchase the edible flowers from the farmers' market or clip from your own garden. Prepare them by rinsing and then removing blossoms and leaves from their stems.

2. Place a generous handful (or more, to taste) of chopped herbs and flowers with the eggs in a bowl.

3. Using a fork, whip the mixture until it is frothy. Add a dash of milk or cream for added richness, if you like. Gently stir in the salt and pepper.Melt the butter in a skillet over medium heat. Pour the egg mixture into the pan and let cook without disturbing it until the top of the frittata has a thin skin on it; then flip and cook the other side, 5–7 minutes total.

4. Mix the yogurt with chili oil to taste, depending on how spicy you like your food. Slather the yogurt mixture on the eggs, and then decorate the top with whole flowers and petals. Serve warm or at room temperature.

In place of flowers, a generous helping of herbs surprises and delights.

TUREEN OF HERBS

Keep a soup tureen filled with an abundance of herbs on the kitchen counter and it's easy to spontaneously sprinkle bits of parsley and arugula flowers on every meal.

Soup tureens often sit in kitchens unused, and I liked the idea of filling one with assorted herbs and edible flowers. Keep it close to where you are cooking and clip from it whenever you need to add flowers and herbs to dishes.

When arranging the bunches of flowers and herbs, layer them, keeping them grouped by type initially. It's nice to have the herbs as organized and bunched together as possible so that you can grab what you need without having to search it out. These should last around a week depending on the temperature of your room. Whatever you don't use can be hung upside down to dry for use later.

You will need

- About 45 stems in total of mixed herbs
- This bowl includes basil, bay leaves, garlic flowers, arugula and arugula flowers, lavender, and rosemary.
- A soup tureen

In place of flowers, a generous helping of herbs surprises and delights.

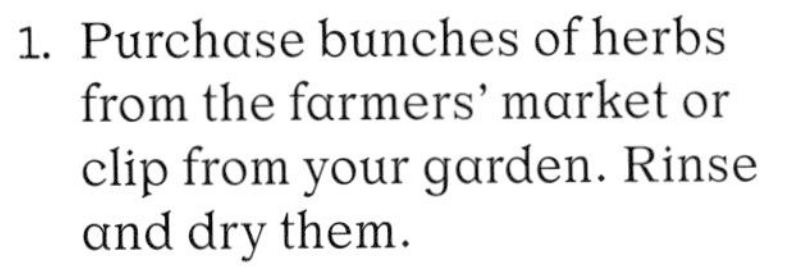

1. Purchase bunches of herbs from the farmers' market or clip from your garden. Rinse and dry them.
2. Arrange the stems in clumps in the tureen, leaning them on one side to start.

3. After you have a bit of structure in the bottom of the dish you can add individual stems, which will add more movement to the arrangement and help the structure of the arrangement by intertwining the stems further.
4. Add water to the tureen and check the water level regularly (as this is a shallow style of bowl, it will need frequent refilling).

chapter 3 /

Bedroom

Flowers and herbs can be an essential part of every sleeping ritual. Whether it's a bowl of dried petals next to the bed, a garland of herbs to induce sleep, or a bedside arrangement, floral elements can provide rest and comfort for this quiet part of the day. One of my favorite memories from my travels is when I rented an apartment that had a mattress on the floor. I went to the flower market and purchased a tall stem of lily and placed it in a jar on the floor next to the bed. Though the scent was a bit overwhelming, sleeping under a lily blossom was a romantic way to embrace the sparsely decorated room.

ROSE AND PEONY POTPOURRI

Forget the potpourri you remember from the 1980s. You can make this simple, modern version with whole flowers and petals, a couple of herbs, and a single essential oil. It's a fabulous winter indulgence to remind you of the roses of summer all season long.

Potpourri has always been one of my favorite flower crafts to make. As a teen, I spent hours in the kitchen concocting flower mixtures while studying herbal books. A trip to the herb store in downtown Olympia was always a treat. I would scoop bags full of dried chamomile, verbena, rose buds, and rose petals and then come home and make bowls of potpourri for gifts or to spread around the house in containers and jars.

Though many of us have a 1980s reference for potpourri in America, records of it go all the way back to the Middle Ages. The word is French for "rotten pot," and the pots of dried flower petals, herbs, and spices were a way to keep odors at bay.

This is my modernized version of potpourri. Instead of the wood chips often used in decades past, it relies on orris root, which is a mixture of two types of iris root, as a natural fixative to help retain the fragrance. Orris root can be a bit tough to find locally, but I was able to secure some from an old-fashioned apothecary where they brought a box out from the back of the shop. With a bit of research you should be able to find it at your local herb shops in the United States, or you can order it online.

Often when I make potpourris, I add different herbs and spices, but I love the visual simplicity of whole flowers and petals. In this case, I've focused on roses and a few peony petals, which I air-dried. The garden roses are amazingly fragrant all on their own, and as you make this, the room will fill with the fragrance of roses. Who doesn't love a pure rose fragrance?

You will need

- 10 stems of garden roses
- 3 stems of peonies
- Rose essential oil
- 3 tablespoons orris root
- A black or dark-colored apothecary jar

Dried blossoms let you add the heady scent of summer to any room—in any season.

1. Garden roses, even from the florist, will work better than hybrid tea varieties that don't have a strong fragrance. Pick them in the hot sun while the oils are at their peak. Alternatively, purchase the flowers from a market, or skip ahead and purchase dried rose petals in bulk online or from an herb store.
2. If working with fresh flowers, you'll need to dry them (see page 86 for details on drying flowers). You can dry the petals on a screen, but if you live in a small space, hanging them upside-down can be easier. It's best to do this in a cool, dry, dark space if possible. Too much light and sun will fade them.
3. Once the flowers are fully dry, remove all the petals, separating out and discarding any rotten or moldy petals. If any petals aren't completely dry, you can put them out in the sun for an hour or so to crisp them up.

4. In a bowl, place whole flowers and petals, along with a few leaves for color, being sure to mix only completely dried ingredients together. In a separate bowl, combine the orris root with about ten drops essential oil, and mix well.
5. Toss the orris root mixture with the petals, leaves, and flowers, mixing well.
6. You can edit the final mix to make sure there is a balance of color and texture. Let the potpourri meld in a jar for 6 weeks; a black apothecary jar will keep the light out and retain color and fragrance. (You can skip this step, but the potpourri will fade sooner.) Afterward, you can put it in bowls to display.

SWEET DREAMS GARLAND

For your bedroom or a guest room, a garland of sleep-inducing herbs such as lavender and geranium is a thoughtful and imaginative touch.

Making my bedroom into a sleep sanctuary creates a grounding way to start and end each day. Given the effect of screens and long days, being intentional about sleep can be transforming.

I keep a bowl of lavender next to my bed. It was harvested from my parents' small farm, so every night I think of my mom's lavender garden and the sweet, calming fragrance helps me get a good night's rest. That's where I came up with this idea for a sleeping garland. I love the idea of this whimsical decoration, to infuse your bedroom with herbs that assist with sleep and also provide an earthy, visual ornament to gaze at each morning and evening. The garland naturally dries and will last about six months, so it will far outlast a vase of flowers. And it's compostable and biodegradable, made entirely with natural materials and no wire.

This version is made with vines for the base, and stems of lavender and scented geranium. Together they make a powerful sleep aid. Olive greens and branches add movement and visual interest. You could reinterpret this recipe with chamomile or lavender flowers, or whatever suits your own taste and supplies.

You will need

- A few long vines like ivy, Virginia creeper, or grapevine (about 3 branches, each in a length of at least 12 to 24 inches (30 to 60 cm) for a single bed)
- 10 stems scented geranium
- 10 stems olive leaves
- 30 stems lavender foliage
- All-natural twine

A winding length of scented herbs makes for deep sleep and sweet dreams.

A garland brightens a
bedroom in any season.

A winding length of scented herbs makes for deep sleep and sweet dreams.

1. Gather the herbs, greens, and vines from your garden, a friend's garden, and/or your foraging spots. Once home, remove the leaves from the bottom part of the herb and greens stems so you'll be able to secure them to the vine neatly. Remove all the leaves from the vines.

2. Make small bundles with the herbs and greens, securing them with twine. Tie a long piece of twine to the top of one of the vines and wrap each bundle onto the vine with the twine, pulling it tightly at an angle. As you add each bundle, overlap the previous stems with the top of each new bundle.

3. Incorporate the next large vine to extend your garland by adding it into one of the bundles as you attach it. Wrap the last section tightly and then tie it off with another piece of twine.

4. Wrap additional vines or olive branches around the garland, weaving in the stems to secure them. You can also weave in small stems in areas that need to be filled in. Adding in these vines will make the garland more stable and add a nice sense of movement.

QUEEN ANNE'S LACE AND JASMINE FOR A BEDROOM DRESSER

A tall, ethereal arrangement of foraged blooms can bring calm to a cluttered bedroom space.

Often, I find my bedroom dresser becomes cluttered with all sorts of small items. To counteract that, I like to put a big, tall arrangement of flowers on the top. I collect modern ceramic vessels for spontaneous foraged finds. This vase from Brazilian ceramicist Mada Pereira, whose daughter owns a shop near my home, looks ethereal when filled with Queen Anne's lace.

I also added a few stems of jasmine, but be careful not to overdo it with this heady bloom: I made myself sick one time filling my bedroom with dozens of jasmine stems. The fragrance can send even the most ardent flower-lover like myself into a day-long migraine and stomachache.

Feel free to play with proportions with an arrangement like this one, going beyond the floral designer's traditional rule of one and a half to two times the height of the vase. I like to go to two and half or even three times the height, using light and airy ingredients to add a sense of visual breathing room and extra height.

You will need

- About 30 stems foraged Queen Anne's lace and other wildflowers and grasses
- The bouquet shown also includes thistle, meadow grasses, honeysuckle, and jasmine.
- A tall vase or pitcher in a barrel shape

Summer wildflowers beckon for a spontaneous flower composition (opposite).

1. Gather your blooms and grasses from a garden or your local foraging spots. Here, the blooms are honeysuckle, jasmine, thistle, and Queen Anne's lace. Once home, trim the stems and, wearing gloves, strip the thorns from the thistles if using.

2. Arrange the Queen Anne's lace in the vase first, as its long stems and umbel flowers are the showstoppers of this arrangement. Clip them to different heights, some long and some short, to let your eye focus on each bloom.

3. Next, add your thistles, then tuck in the wispy stems of honeysuckle, jasmine, and meadow grasses near the rim of the vase to add some curving lines and contrast.

BEDSIDE WISTERIA AND CLIMBING ROSES

Re-create this unabashedly romantic arrangement of roses and other garden blooms and vines in a drinking glass, and it will make your bedroom feel abundant with fragrance and femininity.

At the end of a long day, I find great solace in taking the time for a simple bedtime routine: making the bed with clean sheets, doing a quick five-minute tidying, and adding a bouquet of flowers next to my bed to induce sweet dreams. This routine sometimes feels luxurious on a busy day when I find it difficult to slow down, but it is also a practice in learning to want what I already have.

Sometimes I'll simply pop out to my terrace to clip a stem for my bedside. But other days I want something more indulgent, like this arrangement. These stems are quite obviously not grown for the floral industry, which makes them even more charming. Their homegrown characteristics include a less-than-straight stem, blossoms that nod down, leggy vines with no leaves where they've grown in the shade, and brown petals here and there.

These roses came from a neighbor. I asked if I could clip and cut a few stems of roses and wisteria as I saw her in her yard pruning. I offered to pay for them, but she gave them all to me free of charge—and added an armful of rosemary stems, too!

You will need

- 3 stems Cécile Brunner, or similar climbing roses
- 3 stems privet
- 5 stems abelia
- 5 stems honeysuckle
- 5 stems wisteria

A summer mix of herbs and flowers chosen for their lavish scents.

1. Look for homegrown blooms—in your own garden or that of a neighbor (and ask politely if you can pay for a few). Trim the stems and prepare them by removing the leaves from the bottom of the stems and conditioning them (see page 80) in water before arranging.

2. Add water to a tall drinking glass, then start placing stems. Begin with stems of privet and abelia and then add the long vines of roses.

3. Alternate honeysuckle and wisteria until the arrangement is stable and has a shape you like. Because the glass is narrow, you can add multiple vines and still have a stable arrangement, which creates a lot of movement, fragrance, and softness.

GUEST ROOM DAISIES

Greet a guest with a cheerful little posy of dandelions, daisies, and grasses in a handmade ceramic cup.

Placed by a bedside, a little daisy-and-dandelion bouquet is both discreet and colorfully uplifting. These Spanish daisies grow wild along all of the stone walls in Sintra, but you probably have a version where you live, too. At my home in Portland, Oregon, it's the English daisy that pervades lawns in the spring, and it's always a sign that warm weather is on its way. Any type of daisy-like flower from chamomile to feverfew would work for this arrangement.

You will need

- Spanish daisies, or other daisy-like flowers
- Dandelions
- Grasses
- Small ceramic cup or jar

Bright-eyed daisies to welcome your guests.

1. Gather your blooms and grasses from your lawn or by the roadside.

2. Arrange the flowers in your hand, placing some of the dandelions and grasses a bit higher than the daisies. Add water to the jar, cut the stems of the bundle, and drop it in. Add more stems or adjust as necessary.

3. Check the water frequently—especially important for small vessels like this.

chapter 4 /

Bath

I like to keep flowers in the bathroom low-key, but thoughtful. Stay away from dried blooms in this space as they will not last well with the humidity. Fresh flowers, on the other hand, will love it. Creating canopies with fresh green branches, floating fragrant flowers in bowls of water, and placing a few aromatic stems by the sink will uplift your daily rituals. You might be surprised by how much they truly make a difference in this utilitarian space. I've transformed my own, less than ideal, bathroom interiors with freshly laundered towels, baskets, hanging branches, and a few fragrant flowers.

SCENTED FLOWER MEDITATIONS

Sprinkle flower blossoms into ceramic bowls to bring perfume and beauty to the bath.

Even a single, fragrant, floating bloom can infuse a washroom with delight! Set a bloom, or a few small ones, afloat in a bowl of water near the sink or on a windowsill, or even carry one to the bathtub to gaze at during a meditative bath time. I love to take a little bowl of flowers to the edge of the bath for an at-home spa escape.

You will need

- 1 flower, or several
- The flowers shown here are sweet William and apricot garden rose. Other good choices include tuberose, hyacinth, and orange blossom.
- 1 or more small ceramic bowls

1. Snip a stem, or a few, from a garden or purchase at a farmers' market or wholesale flower market. Once home, clip the flowers from their stems so they will float. Or you can just pinch them off with your fingers if you like; I normally do as I love to absorb a bit of their fragrance and enjoy the tactile experience. A simple gesture like this can turn a morning around!

2. Float several sweet William blooms in one bowl and/or display a rose on its own. Using handmade ceramics with interesting textures or shapes adds interest.

EUCALYPTUS SHOWER SWAG

A few eucalyptus stems hanging from the shower spray can be a charming surprise and a lovely way to start the day.

For those of us who dread mornings, this swag of eucalyptus will fill the shower with fragrance and might just be the lift we need. Use the blossoms as well as the leaves, if you can find them, as they add the slightest floral touch and bit of color. Three different types of eucalyptus are featured here: Small leaf parvifolia, red gum, and true blue. Including a few varieties like this adds visual interest with different textures, colors, and a variety of fragrances. A swag like this one should last a couple of weeks in the shower.

You will need

- 1 dozen eucalyptus stems, in multiple varieties, some with flowers if possible
- Florist's twine, for binding and hanging

1. Purchase the eucalyptus at a flower market. Before arranging, remove the leaves from the bottom portion of the stems.

2. Start by laying the longer stems on the ground for easy composition and to be able to see the full design. Layer medium-length stems over them, and lastly, add the flowers and shorter stems. Finish by adding in a few random stems here and there, alternating the varieties of eucalyptus, so that it is isn't perfectly layered.

3. Bind the bundle securely with twine and hang from the showerhead.

BATHING BRANCHES

Adding branches in surprising spots can make even the humblest of washrooms feel lively and lush.

No matter how tiny or large your bathroom, you can give it new life and personality with the simplest of natural ingredients: branches. I used olive branches here, hanging them from their natural little crooks right on the towel racks, and added a big bundle in a large vase. You can set a tall display on a chair, or even on the floor—I love floor vases and they work for anything that is a bit larger without using up too much space in a small room. Decorating a washroom with branches like this is an easy way to bring a low-maintenance bit of nature indoors. They will love the humidity of the bathroom. Be sure to change out the water after a week. After two weeks you can replace with fresh stems. The stems that are out of water will begin drying, and you can transfer them to a less humid room later to complete the drying process.

You will need

- A variety of small branches, such as eucalyptus, bay, or olive
- A large vase

Olive branches on a towel rack.

Olive branches in a cottage bathroom.

CHAMOMILE AND ELDERFLOWER FACIAL STEAM

Enjoying a facial steam of elderflower and chamomile—from your own garden, or the farmers' market—is an easy and inexpensive way to treat yourself to a few moments of silent relaxation in your day.

As a teenager I spent most of my time growing a cottage garden. I am drawn to the idea of a *horta*, as the Portuguese call this kind of garden, because it is a mix of everything beautiful and useful. Basically, you can grow anything in the little plots, from ornamental flowers to vegetables. In the summers, I would use the chamomile I grew for teas, hair rinses, and facial steams. Its restful qualities always calmed me down.

You can do the same, whether or not you cultivate your own garden plot (just use flowers from a clean, chemical-free source). Take time for your skin and mind as the sweetly scented elderflower and chamomile steam opens your pores and forces you to focus on breathing for a few minutes.

You will need

- 1 handful of chamomile flowers
- 2 flower heads of elderflower
- A bowl for steaming

Elderflower and chamomile add a sweet take on a facial steam.

1. Gather fresh chamomile and elderflower from a garden, or the farmers' market. Rinse quickly in cold water.

2. Take a handful of elderflower and a handful of chamomile flowers (you will need approximately the amount pictured at left).

3. Place everything together in a bowl. Pour boiling water over the flowers and let them steep. Wait a few minutes so as to not expose your skin to too much heat. Grab a towel and drape it over the bowl, then duck your head under the towel to catch the steam. Breathe slowly. Stay there for three to five minutes. Dry your face, then rinse with cold water to close the pores and dry again.

chapter 5 /

Garden and Terrace

You don't have to have a garden to infuse your outdoor space with flowers in myriad ways. And adding decorative floral touches to the outside of the home needn't be just for holidays or special occasions. It can be a way of appreciating the natural abundance around you on an everyday basis. Taking the time to make a simple centerpiece for a gathering, to weave flower crowns with friends, or serve a flower-infused lunch on the terrace are all accessible ways to do that, and they don't have to be formal. And of course, you can grow flowers for cutting in the smallest of spaces—even in the city!

BALCONY CUTTING GARDEN

Why should country dwellers be the only ones who can cut a little bouquet to bring inside whenever the mood strikes? A cutting garden in the city will bring ongoing joy!

Even if you have no backyard, you can make it easier to bring flowers inside by creating a miniature cutting garden on your terrace or balcony. In summer, many annuals and perennials will bloom all season, as long as you keep cutting them. And as they grow, you can divide them into other containers or repot them to give away as gifts.

You don't need a lot of space! Find a plot or area with sun. Partial sun is ok. Even if it's a bit on the shadier side, it just means your flowers will have more foliage. If you live in a cold climate, start the seeds inside in the fall; you will need to transplant them outdoors later. Alternatively, you can plant them directly into the outside containers in the spring, after the threat of frost is over. If you live in a mild climate where it does not get below freezing, you can plant the seeds in the fall right into the outdoor pots, for blooming in spring or summer. Starting the seeds early can encourage longer stems and stronger flowers.

When you are harvesting from your own garden, pick the flowers before the pollen has emerged; if it has, they just won't last as long. Be sure to keep picking the flowers to encourage more growth and more blooms!

A LIST OF FAVORITES

- Bachelor's buttons
- Cosmos
- Cleome
- Gaura
- Marigolds
- Nasturtium
- Sweet peas
- Pansies
- Tomato branches
- Chive blossoms

You will need

- A selection of plants (see my list of favorites above), either purchased from a nursery or grown from seed
- Others that you can purchase in plant form for continuous blooms include rudbeckia, coreopsis, Russian sage, lavender, coneflower, scented geranium, pansy, sage, and rosemary.
- A selection of pots with drainage holes
- Organic potting soil
- Spray bottle of water

Lush gaura grows in a city cutting garden.

1. Choose flowers of different heights and scales, and pots in one color (or just a few) for the most visually pleasing container cutting garden.

2. Before potting the plants, review the basics: Your pots must have drainage! Keep small stones in the bottoms of them and make sure there is a hole for water to escape. Also, use good organic potting soil from your local nursery. If you are starting the flowers from seed, toss the seeds into the pot, and keep them consistently moist—but not soggy—with a spray bottle of water. As they develop, you can thin them a bit to give them space to grow by removing some of the seedlings that are too dense. Keep seedlings in a shady area to start, and follow direction from the seed packets on where the particular varieties you choose will do best. Flowers like cosmos can thrive in full and part shade and you can move the pot around accordingly.

MEDITERRANEAN-INSPIRED FLOWER CROWN

Invoke A Midsummer Night's Dream at your next party by crafting crowns of blooms.

Flower crowns are a summer ritual in some parts of the world, and they can be a great activity with friends or kids, at a picnic or just about anywhere! I like the Mediterranean-inspired feeling of this crown of eucalyptus and sedum. It's light, airy, and dainty, and the copper wire shimmers romantically in the sun.

My friend Gemma wears a flower crown at Lisbon's eighteenth-century secret garden, Jardim da Tapada das Necessidades.

You will need

- 10 stems of small-leaf eucalyptus
- 10 stems of sedum flowers
- Florist's twine
- Copper wire

How-to

1. Purchase the blooms at your local flower market. At home, make small bundles with the greens and flowers and tie them with twine.

2. Make a circle with the eucalyptus, removing about half of the leaves, and securing it by weaving it around itself. Use additional pieces of eucalyptus to make it secure, weaving them around. You'll need to use the most malleable of stems for this.

3. Attach the bundles to the crown with copper wire. Secure them to the eucalyptus base by tightly wrapping the wire at an angle where they are secured with twine. Adjust the greenery or trim as needed.

APPLE PICNIC CENTERPIECE

Anchoring a family-style table, a display of branches laden with fruit elevates a simple picnic to a special occasion.

At an outdoor celebration, a larger centerpiece can feel festive. I like to style with terra-cotta vases (like the thrift-shop find shown on the next spread). There's no need to buy new vases as there are so many to be found at antique stores, garage sales, estate sales, and vintage shops. Scour your local thrift shops for interesting shapes and finishes.

I built the arrangement around these charming apples, still on their branches, from my local flower market. And I tucked in a single stem of foliage from the bauhinia tree: Its leaves have a rounded shape and a chartreuse color that fill in the space to make everything look a bit more lush. I like that the green of the leaves matches the apples, so as not to create any visual clutter.

Whenever you're working with woody stems like this, be sure to use the method for cutting them that we cover in the Basics chapter (see page 82). That will help them retain their fresh look. This apple display serves as a sort of canopy over the picnic without feeling overdone.

You will need

- 3 stems of small, young apples
- 1 stem of bauhinia
- Large terra-cotta vase
- Stones for bottom of vase

1. Gather the apple branches from a tree if possible, or from your flower market, along with the bauhinia. On your work surface, prep the woody stems using the technique described on page 82.

2. Add stones to the bottom of the vase to help hold the stems in place.

3. Start arranging with one tall apple branch, and then add the others in at varying heights, creating nice arches and curves, and cutting one branch lower so the apples hang at the rim of the vase.

4. Place one branch of foliage from a bauhinia tree to fill out the arrangement.

Bauhinia branches fill out this arrangement, but you can substitute branches from shrubs and trees domestic to your area.

A handwoven wreath of garden greens hangs on an aged stucco wall in Sintra.

GREEN GARDEN WREATHS

Ivy, plentiful and easy to work with, forms the basis for these leafy outdoor ornaments.

Whether displayed on your garden, terrace, or patio area, or hanging on your front door, these wreaths can add verdancy all year long. They are woven from vines and branches, so wire isn't needed, though you can use a little copper wire to secure them if you want to. In America, ivy is an invasive species, so most public parks are happy to have you cut it! It works superbly for wreath bases as the woody vines are very malleable and dry well.

Fill out your wreath by adding in foraged greens or cuttings from your garden. You can spray the greenery with water to keep it fresh when hanging on your front door, but if you live in a rainy climate it should last for at least a month outside.

You will need

- 10 ivy vines, each about 12 inches (30 cm) or more in length
- About 25 stems of foraged greens
- Copper wire, optional

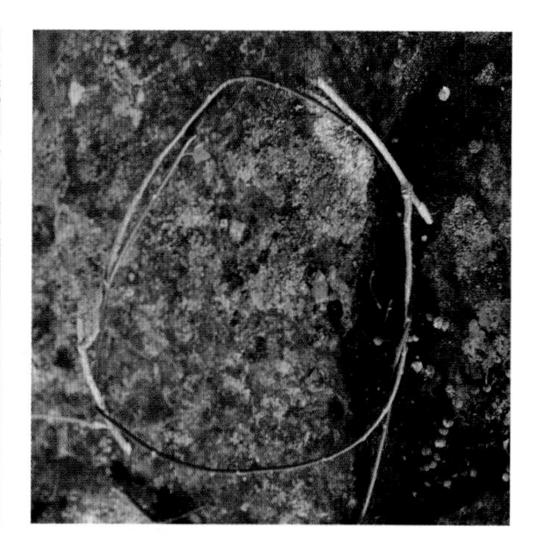

1. Forage for ivy and other greens on your property or in a local park or other foraging spot. Look for sections of ivy that are sturdy, green, and bendable.

2. Remove the leaves from five or six ivy vines.

3. Next, pick out a sturdy but malleable length of ivy and form it into a circle. Wrap it over itself. Then weave a second ivy vine without leaves around the first. Continue adding ivy until the wreath feels secure and you have the thickness you'd like (below).

4. Weave some ivy with leaves or other foraged greens into the base. To further secure the wreath after you've woven in the branches, you can take a narrower, but sturdy, ivy strand without leaves and wrap it around the wreath. Hang the finished wreath outside on an exposed wall, on a door, or on a terrace.

FLOWER LUNCH FOR TWO

Incorporating flowers into a meal—from the elderflower butter to the nasturtium and calendula salad in this elegant light lunch—can elevate any occasion.

From the Portuguese people, I learned that when it comes to cooking, magic can be made with a few high-quality ingredients, and taking the time to share a few small plates with a friend can turn a day around. Growing a garden and living in a slightly remote area of Sintra, I learned to use what was available to me: the morning bread at the local *vende* (general store) or even my own homemade sourdough; salads made from a few leaves, and edible flowers clipped from the garden. Flowers always elevate my dishes and make them feel completely special even if they are made from frugal circumstances.

I often get visitors from home and need to quickly put together a little meal. I find radishes to be just darling, and so delicious when paired with a good sea salt. I'm sharing a new version of this recipe, created recently after seeing all the elderflowers in bloom. Elderflower butter has just the slightest sweet, floral flavor to contrast the crisp, spicy nature of the radishes.

Another reliable snack that can be dressed up with fresh blossoms is cheese and bread, strewn with confetti flower petals.

And, of course, this meal wouldn't be complete without a flower salad. They were a mainstay of my childhood tea parties and I really have never stopped making them. This salad focuses on the crisp and spicy flavors of calendula and nasturtium paired with greens and herbs. Using herbs as salad greens is a trick I learned from my friend Juliane one summer during my stay on her family's farm in the Brandenburg region of Germany. Each night, her mom walked out to the garden where she gathered dill and parsley to mix in with greens in equal parts. I also notice it in Portugal, where parsley is the star of many dishes. At my local market, the vendors often give away free bouquets of parsley and cilantro with your vegetable purchases.

A meal like this one should be styled with flowers as well. To quickly spruce up a terrace for guests, you can use leftover herbs in vases as decoration around the food. It's easy and adds a bit of life to the table. You can also green things up by adding herbs in pots and even stringing up dried herbs.

For flowers, try a simple mix of blooms purchased at the flower market—like the bouquet at left of peonies, scabiosa, feverfew, yarrow, and clematis, which I set in a colorful, handmade pitcher. You can arrange most of the flowers in your hand, then add in the clematis vines once you place the bouquet in the pitcher. Then fluff it and adjust as necessary.

ELDERFLOWER BUTTER WITH RADISHES

Use a high-quality butter to create this whipped butter with a subtle floral infusion.

Purchase elderflower and calendula from the flower market or gather them from your garden (just be sure they have not been sprayed). At home, prepare the elderflower for use by shaking and then rinsing the flowers, and letting them sit for a few hours to make sure they are clean, and any bugs have been removed. Next, remove the elderflower blossoms from the stems. This is necessary because the stems and leaves of elderflower are toxic to consume.

Pour the boiling water over the blossoms and let them sit for 24 hours. Keep the mixture in the refrigerator once it cools down.

The next day, bring both the butter and the elderflower infusion to room temperature, then stir 1 tablespoon of the elderflower infusion into the butter, and mix well. Form the butter into a roll, roll it in calendula petals and then wrap in parchment paper. Chill to firm up for at least a few hours. You can keep the rest of the infusion in the fridge to add to drinks. It will last a few days.

Serve with fresh radishes and a little salt.

NOTE: Never eat elderflower stems, leaves, or whole blossoms, and do not use them as a garnish. The elderflower petals are not edible raw; only use the petals to make the infusion as described, but do not use any parts of the rest of the plant.

You will need

- 2 large stems of elderflowers
- 1½ cups (360 ml) boiling water
- 1 cup (2 sticks) salted butter
- 5 blossoms of quantity calendula from the farmers' market or your own garden. Rinse quickly in cold water. Let dry. Remove the petals and discard the rest of the flower.
- Parchment paper
- Radishes
- Sea salt

NASTURTIUM, HERBS, AND CALENDULA WITH PARSLEY DRESSING

This salad focuses on the bright colors and spicy taste of nasturtium flowers and leaves, accented with calendula petals, refreshing mint, basil, and parsley.

Pick from your garden or purchase from a farmers' market a bundle each of mint, basil, and nasturtium (and parsely for the dressing). Separate stems and remove old leaves. Rinse and dry. To make the salad, remove the mint and basil leaves and toss together, adding the nasturtium flowers last.

FOR DRESSING:
Mince parsley leaves, then mix with the olive oil, Dijon mustard, and salt and pepper.

You will need

- A bundle each of mint, basil, nasturtium, and parsley
- ¼ cup (60 ml) olive oil
- 1 teaspoon Dijon mustard
- Salt and pepper, to taste

PANSY AND THYME FLOWER CHEESE

Paired with rustic bread, this little appetizer feels fancy and fun.

Purchase the pansies and thyme at a flower or farmers' market, or gather them from your garden, and prepare them by rinsing lightly and patting dry.

Remove the pansy petals and place them on a plate, then sprinkle them with thyme leaves. Next, roll the fresh cheese in the mixture until coated. Serve with a generous splash of olive oil or honey, and slices of bread.

You will need

- 5 to 10 stems of pansies
- 1 bundle of thyme
- 1 wheel of goat cheese (8 oz. / 227 g) or another fresh, wet, or sticky cheese
- Olive oil or honey, for serving
- Rustic bread

Index of Flowers

The common names of flowers included in this book:

Field, Flower, Vase Sketchbook

As I was putting this book together, my friend Rita Teles Garcia often visited—and sometimes she made a sketch of that day's project. I thought these were so irresistible, I wanted to include them here for you to enjoy.

page 161

Warm Plums with Lavender-Infused Mascarpone and Thyme Blossoms

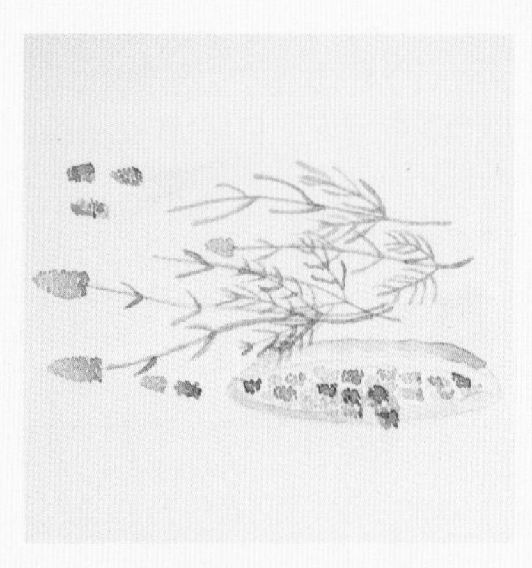

Prepare the lavender by removing the blossoms from the stems.

Gently simmer the blossoms in the cream.

Let cool, then strain the blossoms out.

Heat the pitted and halved plums in butter until just warm.

Plate the mascarpone, top with plums, and drizzle with honey, and serve.

CERAMICS

Ceramics and textiles in this book are from the following Lisbon-based designers:

MARIA CASTEL-BRANCO
Travessa Teixeira Júnior 25, 1300-553 Lisbon
mariacastelbranco.pt

CECILE MASTELAN
Calçada da Estrela 3 5, 1200-661 Lisbon
cecilemestelan.com

MADA PEREIRA AT D'OLIVAL
Rua Poiais de São Bento 81, 1200-347, Lisbon
dolival.pt

O LEITÃO DO REI
Praça Dom Fernando II 47, 2710-512 Sintra

PAUL KOHL AT INA KOELLN
Rua da Silva 27 RC, 1200-446 Lisbon
inakoelln.com

SEDIMENTO / ATELIER E AULAS DE CERAMICA
Travessa Santo Ildefonso 31, 1200-667 Lisbon
sedimento.pt

SAL ATELIER
Sofia Albuquerque
instagram.com/sal_atelier

TEXTILES AND PROPS

BRIC AND BRAC
Rua Alto da Bonita, 1 Sintra

CAVALO DE PAU
Rua de São Bento 164, 1200-821 Lisbon
www.cavalodepau.pt

ISABEL LARA DESIGN EM PANO LDA
Rua do Poço dos Negros 181, 1200-338, Lisbon
instagram.com/isabellaradesignempano

VINTAGE AND TERRA-COTTA VASES

A VIDA PORTUGUESA
Rua Anchieta 11, 1200-023 Chiado, Lisbon
avidaportuguesa.com

LUIS LOPES
Rua de São Bento 266, 1200-817 Lisbon

EDIBLE FLOWERS

MERCADO BIOLÓGICO DO PRÍNCIPE REAL
Príncipe Real, Lisbon

FLOWERS

ALCINA AT MERCADO RIBEIRA
Avenida 24 de Julho s/n, 1200-481 Lisbon
Instagram.com/alcina_das_flores

FORAGED IN SINTRA
Colares, Penedo, Portugal

FLORISTA CAMÉLIA

MERCADO ORGANICO PRÍNCIPE REAL

MERCADO DO ALMOCEGEME

PHOTOGRAPHY LOCATION IN SINTRA

RIBEIRA DA SINTRA
Sao Pedro da Sintra, Penedo, Colares, Lisbon (Santos, Mouraria, Sao Bento)

PHOTOGRAPHY TAKEN AT THESE SINTRA COTTAGES AND LISBON HOMES AND STUDIOS

CASA NAS NUVENS

CASA DO CASEIRO

URBAN VILLA

SECOND HOUSE, LAPA

INA KOELLN

O CORVO CAFÉ

LIVING SIMPLY LISBON

MY FAVORITE LISBON FLOWER SHOPS

PEQUENO JARDIM
pequenojardim.com

MAGDALA FLORES
instagram.com/magdalaflores

CALLAS
Rua de Sant'Ana à Lapa 69A, 1200-798 Lisboa

SAUDADE FLOWERS
saudadeflores.com

ALCINA AND THERESA AT
Mercado Ribeiera Avenida 24 de Julho, Lisboa

KCKLIKO
kckliko.com

FLORISTA RUA FERREIRA BORGES
Rua Ferreira Borges, Campo D'Ourique, Lisbon

LISBON AND SINTRA MARKETS

ALMOCEGEME FERIA

SINTRA MERCADO

MERCADO DO ORGANICO PRÍNCIPE REAL

FLOWER WHOLESALERS IN THE USA

LOS ANGELES FLOWER MARKET
754 Wall St., Los Angeles, CA 90014

SAN FRANCISCO FLOWER MART
640 Brannan St., San Francisco, CA 94107

OREGON FLOWER GROWERS ASSOCIATION
3624 N Leverman St., #3945, Portland, OR 97217

TWIN CITIES FLOWER EXCHANGE
1790 Larpenteur Avenue West, Falcon Heights, MN 55113

MIAMI FLOWER MARKET
6964 NW 50th St., Miami, FL 33166

NEW YORK FLOWER MARKET
128 W 28th St., New York, NY 10001

CUT FLOWER WHOLESALE
2122 Faulkner Rd. NE, Atlanta, GA 30324

SEATTLE WHOLESALE GROWERS MARKET COOPERATIVE
665 S Orcas St., Seattle, WA 98108

SLOWFLOWERS.COM
Slow Flowers network of sustainable flower growers

ASCFG.ORG
Association of Cut Flower Growers
Peterkort Roses, Portland, OR

FOR FLOWER ID

WILDFLOWERSEARCH.ORG

UMA MAO CHEIA DE PLANTAS QUE CURAM BY FERNANDA BOTEHLO ASSISTED ME WITH WILDFLOWER ID IN SINTRA.

SAFETY

The material contained in this book is presented only for informational and artistic purposes. If you use plants or flowers for any of the recipes included in this book we suggest you use only items from farmers' markets or grocery stores. If you choose to eat plants or flowers you may have found in the wild, you are doing so at your own risk. The author has made every effort to provide well-researched, sufficient, and up-to-date information; however, we also urge caution in the use of this information. The publisher and author accept no responsibility or liability for any errors, omissions, or misrepresentations expressed or implied, contained herein, or for any accidents, harmful reactions, or any other specific reactions, injuries, loss, legal consequences, or incidental or consequential damages suffered or incurred by any reader of this book. Readers should seek health and safety advice from physicians and safety professionals.